THE ANALYSIS AND SYNTHESIS OF LINEAR SERVOMECHANISMS

By

Albert C. Hall

Associate Professor of Electrical Engineering

Massachusetts Institute of Technology

A PUBLICATION OF

THE TECHNOLOGY PRESS

MASSACHUSETTS INSTITUTE OF TECHNOLOGY

PRINTED IN THE UNITED STATES OF AMERICA

FOREWORD

This paper is reproduced by the permission of the Massachusetts Institute of Technology to which it originally was submitted as a doctorate thesis. The theory set forth has been tested and proved on numerous servo problems encountered by the Servomechanisms Laboratory at M.I.T. In order to make the development inherently complete, a number of points have been discussed which have been covered previously by other papers. Reference to these papers is made in the text; those of Black,[2,3,] Nyquist,[21,] and Harris,[13,] are worthy of special reference and the paper by Bomberger and Weber,[31,] which was not available to the author until most of the work was completed, is particularly significant.

Chapter II, devoted to the minimum-integral squared error criterion, is included in the paper because of the frequency with which that criterion is applied in servo design. Those who are not specifically interested in that criterion can omit Chapter II without loss of continuity.

Acknowledgment is gratefully made to Professor G. S. Brown, the director of the Servomechanisms Laboratory in which much of the work was done, for his interest and for the fact that it was he who was primarily responsible for arousing interest in servo design; to Professor H. L. Hazen for his continued interest and encouragement; to Professor M. F. Gardner, Professor E. A. Guillemin, Professor J. C. Boyce, and Mr. H. T. Marcy for numerous helpful suggestions; and especially to Mr. G. J. Schwartz for his careful reading of the paper and valuable criticisms.

Albert C. Hall

Cambridge, Massachusetts
May, 1943

SECOND PRINTING

The 1943 printing of "The Analysis and Synthesis of Linear Servomechanisms" was classified as restricted and distributed accordingly. After the close of the war the book was declassified in order to permit general dissemination of the material. It was hoped that it would be possible to revise and add to this material before republication. However, it appears that it will be some time before such revised material appears in the press, and this second printing has been issued to meet an existing need. The second printing is essentially the same as the first, except that the transfer loci have been redrawn and a different method of reproduction employed in order to improve their legibility.

Albert C. Hall

Cambridge, Massachusetts
May, 1947

TABLE OF CONTENTS

CHAPTER		PAGE

THE ANALYSIS AND SYNTHESIS OF LINEAR SERVOMECHANISMS

ABSTRACT

This paper is a formulation of a servomechanism design procedure based primarily upon an analysis of the system response to sinusoidal inputs of various frequencies. Although a knowledge of the transient performance of a servomechanism provides an excellent basis for predicting the response of the system to the conditions of a particular application, the complexity of most physical systems makes the computation of the transient performance and the translation of the resulting information into physical design criteria extremely laborious. On the other hand, knowledge of the response of the servomechanism to sinusoidal inputs is not quite as useful in enabling a prediction to be made of the servo response to the conditions of most practical applications, but the computation of the sinusoidal response and the translation of this information into useful design criteria are much simpler.

The methods developed are based upon the characteristics of the servomechanism transfer-function which is defined as the vector ratio of the servo output to the difference between the servo input and output for sinusoidal inputs of various frequencies. The characteristics of a servomechanism are completely determined once its transfer-function is specified. Moreover, physical devices that realize a prescribed transfer-function are readily synthesized if the given function is of such nature that such devices exist. This paper undertakes first to derive the interconnecting relations between particular servomechanism characteristics and the transfer-function of the servo, and second to synthesize devices that physically realize desired transfer-functions.

The study of transfer-function characteristics is aided by considering the transfer-function to be a vector and plotting the locus described by the tip of the vector as the frequency of the servo input is varied. The application of graphical analysis to this locus, termed a transfer-locus, facilitates the calculation of certain servomechanism characteristics and provides a clearer insight to many servomechanisms phenomena. It is also shown how a knowledge of the transfer loci of a servomechanism enables the system to be so adjusted that optimum performance is obtained.

The performance of a servomechanism may be unacceptable if either the steady-state error or the transient error is unsatisfactory under operating conditions. The steady-state error is defined as the difference (or error) between the servomechanism output and its input under input conditions of such nature that this difference is either constant or varying periodically. The transient error is the difference between the servomechanism output and input under input conditions that result in a nonperiodic variation in that difference. This paper derives the form the servo transfer-function must possess if the steady-state error and the transient error are to fall below allowable limits. It is shown that, if necessary, certain compensating functions can convert the servo transfer-function into the desired form. Criteria are developed that serve as guides in the analysis, application, and adjustment of the devices that physically realize the compensating functions.

Following the discussion of compensating functions, their physical realization is considered and it is shown how several very different types of circuits may be synthesized that yield the desired function. Advantages and shortcomings of the various compensating devices are discussed.

THE ANALYSIS AND SYNTHESIS OF LINEAR SERVOMECHANISMS

INTRODUCTION

It is frequently necessary to control the position of a device or the state of a process in accordance with indications or signals supplied by a suitable controlling instrument. If the power required to operate the device or process is large compared with the power available from the controlling instrument, means must be provided for effectively amplifying the controlling signals in order to secure proper operation of the device or process. The element that amplifies these signals and operates the device or process is known as an automatic controller, and the complete system, comprising the automatic controller, the controlling instrument that provides the automatic controller with signals, and the device or process being controlled, is known as an automatic control system.

Automatic control systems, in general, comprise two types, namely, open-cycle control systems and closed-cycle control systems. The signals supplied to the controller of an open-cycle control system are received solely from the controlling instrument mentioned above, while in the case of the closed-cycle control system additional signals that are proportional to the position of the device or the state of the process being controlled are received by the controller. Use of a closed-cycle control system permits much greater accuracy to be attained than is possible with an open-cycle control system.

A closed-cycle control system is also termed a servomechanism. Hazen[14] has formally defined a servomechanism as "a power-amplifying device in which the amplifier element driving the output is actuated by the difference between the input to the servo and its output." Servomechanisms are used wherever accurate, automatic control is desired. For example, they are employed to steer ships,[17,18] to control airplanes,[1,22] to regulate temperature,[12,27] to maintain liquid levels,[25] to control many military devices, and for general industrial process control.[11,16,23] While a great deal has been published describing particular types and applications of servomechanisms, relatively little has been written concerning the broader aspects of the problem of their analysis and design.

The type of servomechanism discussed in this paper is the type employed when results of highest accuracy are required, namely, the continuous-control type. This type of servomechanism is distinguished by the fact that a definite and continuous corrective action is developed by the servo-controller and applied to the device being controlled no matter how small is the error in the position of that device. This paper is devoted exclusively to the analysis of the continuous-control type of servomechanism.

If the performance of a system can be expressed mathematically by a linear differential equation with constant coefficients, that system is said to be linear. While the following paper is concerned with linear servomechanisms only, the results are guides to predicting the performance of systems with certain nonlinear characteristics. The load of many servomechanisms includes Coulomb friction, i.e., friction whose magnitude is independent of velocity and whose

---

14. Numbers refer to bibliography at end of paper.

direction is such as always to oppose the direction of motion of the servo output. The effect of a small force or torque of this type on an otherwise linear system can be predicted approximately without the need of making an exact analysis of the system, provided the analysis of the otherwise linear system is known. The subsequent theory, however, is devoted entirely to linear systems.

Most of the published works dealing with the subject of automatic control have been concerned principally with the description of the performance and application of a particular automatic control system. Certain earlier papers of analytic nature are listed in the bibliography under items 6, 8, 13, 14 , 15, 18, 19, 28, 29. Further reference is made to these papers in subsequent portions of this paper.

The theory of feedback amplifiers is applicable in many respects to the analysis of servomechanisms and the similarity between feedback amplifiers and servomechanisms is stressed at appropriate points in the paper. Certain important basic differences, however, exist between the two devices. One is that a servomechanism is principally concerned with obtaining high accuracy; much design effort, therefore, is devoted to securing as small a servo error as practicable. On the other hand, most feedback amplifier theory is relatively unconcerned with this problem. A second essential difference arises from the fact that the load on a servomechanism is generally such a comparatively massive member that the important part of the frequency response of the complete system lies in and below the low audio frequency band. On the other hand, most feedback amplifier theory is concerned with securing good response at and above high audio frequencies. The frequency response of servomechanisms can be extended and the servo response improved by employing compensating circuits whose characteristics become unimportant at frequencies higher than the audio range. In most cases, therefore, the effects of parasitic inductances and capacitances present in these compensating circuits are relatively unimportant. On the other hand, the frequency response of most feedback amplifiers is limited by the parasitic inductances and capacitances of the amplifier circuit. The type of compensating circuits applicable in servomechanism controllers may not be employed to extend the frequency range of amplifiers, therefore, because of the effect of the parasitic elements on these circuits at high frequencies. References are made to feedback amplifier theory and portions of the theory are utilized, but, in general, the results in this paper have been developed along separate lines.

Methods of Analysis

The design of electrical equipment of all types is aided by obtaining, by calculation or laboratory measurements, the response of the equipment to certain test conditions so chosen that the resulting effect upon the equipment will yield information pertinent both to its final operation and to improvements in design. In the final analysis, electrical equipment cannot be considered completely tested until operating results under actual working conditions are known. Nevertheless, carefully chosen, easily applied, test conditions will yield information that aids in the prediction of the performance of the equipment "in the field" and serves as a guide to designing improvements in the equipment.

An example of a device which is frequently tested only by obtaining its response to purely artificial conditions is communications equipment. A telephone filter is seldom tested by subjecting it to actual telephone conversations. Instead, sinusoidal voltages of various frequencies are applied to the input of the filter and the amplitude and phase relationship between the input voltage and the resultant output voltage or current are determined. If the frequency range of the input voltage is chosen correctly, the performance of the filter under actual working conditions may be accurately predicted. If the filter is unsatisfactory, the test yields information that is a guide to the design of improvements.

The scientific basis for testing filters with sinusoidal input voltages lies in the fact that if the phase and amplitude response of the filter is known for all input frequencies from zero to infinity, its response may be predicted for any type of input, whether that input is sinusoidal or transient. Mathematically, it is necessary to know only the amplitude or the phase response over the infinite frequency range since the other response may be calculated from the known function. Practically, this calculation is not easy to make, and it is better to determine both responses independently.

More pertinent results may be obtained in certain problems by studying the response of the equipment to transient inputs rather than to sinusoidal inputs. Television amplifier networks are frequently tested in this way, and it has been found that in many cases the results so obtained are more easily interpreted than those obtained by testing the network with sinusoidal input voltages. It may be shown that if the response of a network to any transient is known, its response to all other types of input may be determined. Therefore, any linear system is completely determined if either (1) its amplitude and/or phase response to sinusoidal inputs over the entire frequency range are specified, or (2) its response to some type of transient input is specified.

Whether a particular device is tested by applying a transient or a sinusoidal input is controlled by (1) the ease with which the test may be applied; (2) the closeness with which the test approximates operating conditions; (3) the facility with which the results may be interpreted and transferred into criteria that serve as a guide for further design. In the field of automatic control there has been considerable discussion as to the type of input test that should be applied. In practice a servomechanism may never be subjected to either a sinusoidal input or an instantaneously applied displacement or velocity input. Such test inputs, however, are most frequently used for determining the response of the servo because of the ease with which they are applied and the resulting response measured. While methods have been developed for testing and rating a servomechanism under the actual input conditions to which it will be subjected, these methods at present are used infrequently because either the form of the input is not accurately known, or the labor required to obtain significant results from this type of test is too great.

Hazen,[14] Brown,[6] and others have pointed out the value of determining the response of a servomechanism to a transient input. Briefly, this advantage lies in the fact that the transient input may be so chosen that it closely approximates actual operating input conditions. A knowledge of the transient response, therefore, enables an accurate prediction to be made of the operating response of the system. Although the transient response of a servomechanism, if properly interpreted, will yield almost conclusive results as to whether the servo is suitable or unsuitable for the particular application it is intended, the results obtained from such a test are frequently very difficult to interpret if it is necessary to use the information as a guide to the design of improvements in the system. The reason for this lies in the difficulty attending the transient analysis of all but simple systems.

The response of a servo to its actual operating conditions generally may not be predicted from sinusoidal test data as accurately as from transient test data; however, sinusoidal test data, if correctly interpreted, yield information that is much more valuable than the transient test data if it is desired to improve the servo performance. Obviously, the two types of tests can be made to complement one another.

This paper endeavors to show how sinusoidal analyses may expedite the design of servomechanisms. Throughout the paper, those factors are emphasized that are important in obtaining good transient response, since in the final analysis the servomechanism must respond well to inputs of transient nature. A procedure that has been found very satisfactory is to (1) carry out the design on a sinusoidal basis; (2) set up the system in the laboratory, using the design results; and (3) make such a final adjustment of the parameters that optimum transient response of the system is attained. The final adjustment makes allowance for the fact that many parameters important in servo design are difficult to predict with high accuracy.

The synthesis of compensating devices in this paper approaches the problem as though the devices were always electrical networks. Mechanical, hydraulic, and other types of equivalents exist for these networks, and their use is frequently preferable to that of purely electrical circuits. The theory developed in the following chapters is just as applicable to linear devices of such types as it is to electrical networks.

CHAPTER I

SERVOMECHANISM FUNDAMENTALS

The first chapter is devoted to a presentation of the mathematical reasons for employing a closed-cycle control system and a summary of servomechanism fundamentals. The mathematical methods utilized throughout the paper are also introduced.

<u>Advantages of a Close -Cycle Control System</u>

The reasons for employing a closed-cycle control system are best explained if the conditions necessary to secure high accuracy operation with such a system are compared with the conditions that must be maintained if an open-cycle control system is to give corresponding accuracy. Accuracy is the primary requirement upon an automatic control system and the following analysis makes clear the advantages of a closed-cycle system over an open-cycle system in this respect. A block diagram of an open-cycle system is illustrated by figure 1. The input to the

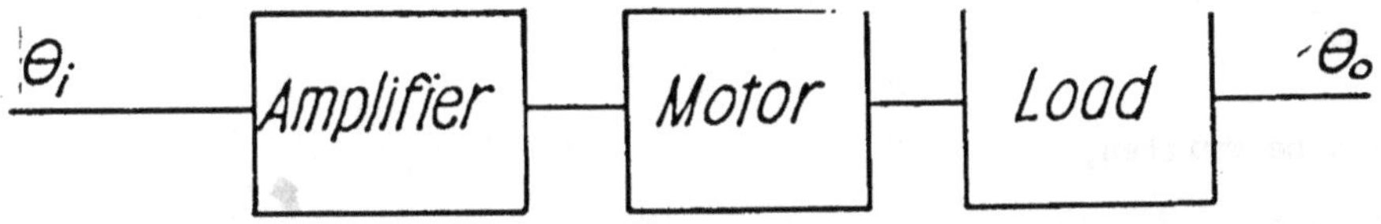

FIGURE 1

system is expressed as a function of time by $\theta_i(t)$ and the output as a function of time is represented by $\theta_o(t)$. $\theta_i(t)$ and $\theta_o(t)$ may represent a linear or angular position, a voltage or a current. The response of the open-cycle system is completely specified if the relation between $\theta_o(t)$ and $\theta_i(t)$ is fixed.

If $\theta_i(t)$ is a sinusoidal function of a particular frequency, phase, and magnitude, the response of the controller may be specified by prescribing the relationship by means of which the phase and magnitude of the corresponding sinusoidal output may be calculated. Thus if

$$\theta_i(t) = A_i \cos(\omega t + \phi_i),$$

the output may be expressed as

$$\theta_o(t) = A_o \cos(\omega t + \phi_o).$$

The relations between the magnitudes A_i and A_o and the phases ϕ_i and ϕ_o specify the response the control system.

Equations (1) and (2) may be written in another way:

$$\theta_i(t) = \mathrm{Re}\left[A_i e^{j\phi_i} e^{j\omega t}\right]$$

$$\theta_o(t) = \mathrm{Re}\left[A_o e^{j\phi_o} e^{j\omega t}\right]$$

The symbol $\text{Re}[\ \ ]$ signifies "the real part of." The quantities $A_i e^{j\phi_i}$ and $A_o e^{j\phi_o}$ are vectors and the ratio of the second to the first is also a vector. The response of the control system is completely prescribed by specifying this ratio which is known as the transfer-function of the system and is represented by $P(j\omega)$. The magnitude and the phase of the vector ratio depend upon the frequency, ω, of the input to the system.

$$P(j\omega) = \frac{A_o e^{j\phi_o}}{A_i e^{j\phi_i}} \tag{5}$$

The functions $A_i e^{j\phi_i}$ and $A_o e^{j\phi_o}$ are frequently denoted by the symbols $\Theta_i(j\omega)$ and $\Theta_o(j\omega)$. Thus

$$\Theta_i(j\omega) = A_i e^{j\phi_i}, \tag{6}$$

$$\Theta_o(j\omega) = A_o e^{j\phi_o}. \tag{7}$$

Equation (5) can be written,

$$P(j\omega) = \frac{\Theta_o(j\omega)}{\Theta_i(j\omega)} \ , \tag{8}$$

Or employing a short-hand used later,

$$P(j\omega) = \frac{\Theta_o}{\Theta_i}(j\omega) \ . \tag{8a}$$

The function $P(j\omega)$ is calculated by methods described in any text on circuit analysis.

Equation (8) expresses the output of the control system as a function of its input provided that the input is a sinusoidal function. An equivalent relation can be written in case $\theta_i(t)$ is a suddenly applied disturbance by employing Laplacian transforms. Suppose $\theta_i(t)$ is some function to be applied to the system and $\Theta_i(s)$ is its Laplacian transform.

$$\mathcal{L}\left[\theta_i(t)\right] = \Theta_i(s) \tag{9}$$

The symbol $\mathcal{L}[\ \ ]$ signifies the "Laplacian transform of" and s is a complex variable. Similarly the transform of the system output is

$$\mathcal{L}\left[\theta_o(t)\right] = \Theta_o(s) \ . \tag{10}$$

The relation between the transforms $\Theta_i(s)$ and $\Theta_o(s)$ for this open-cycle system is

$$\frac{\theta_o(s)}{\theta_i(s)} = P(s). \tag{11}$$

The transfer-function P(s) is the identical function defined by Equation (8) except the frequency variable, $j\omega$, is replaced by the complex variable, s.

The time function, $\theta_o(t)$, caused by an arbitrarily applied input function $\theta_i(t)$ can be calculated by the following steps. (1) Determine the Laplacian transform, $\theta_i(s)$, of $\theta_i(t)$ from a table of transforms or by calculation. (2) Find the transfer-function, P(s), by routine circuit analysis methods. (3) Calculate $\theta_o(s)$ by means of Equation (11). (4) Determine the inverse transform, $\theta_o(t)$, of $\theta_o(s)$ from a table of transforms or by other means.

Laplacian transform theory can be applied to the calculation of the static error of the open-cycle system by arbitrarily applying a change in input position and calculating the output position after a sufficiently long period that steady-state conditions again exist. Assume that the servo input is a time function such as illustrated by Figure 2. The equations for this type of input function are as follows:

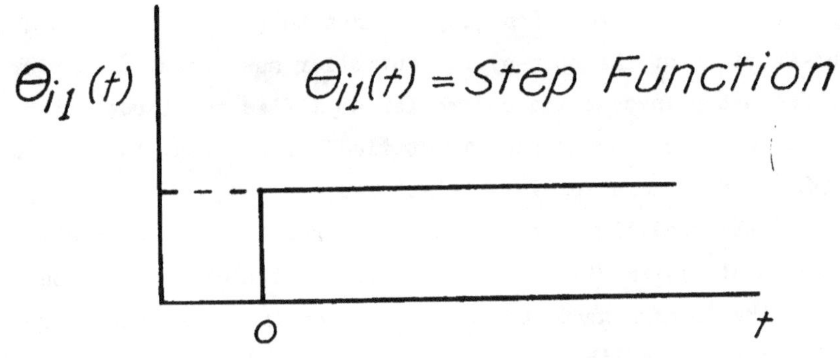

FIGURE 2

$$\left.\begin{aligned} \theta_{i1}(t) &= 0 \qquad t<0 \\ \theta_{i1}(t) &= 1 \qquad t>0 \end{aligned}\right\} \tag{12}$$

This type of input function is known as an unit function and its Laplacian transform is

$$\theta_{i1}(s) = \frac{1}{s}. \tag{13}$$

The Laplacian transform of the output is obtained by applying Equation (11).

$$\theta_{o1}(s) = \frac{P(s)}{s} \tag{14}$$

It is next necessary to calculate the final or steady-state value of the servo output, i.e., $\lim_{t \to \infty} \theta_{11}(t)$.

This calculation is performed very easily by applying the final value theorem of Laplacian transform theory.* This theorem states that if

$$F(s) = \mathcal{L}\left[f(t)\right] \quad , \tag{15}$$

then in almost every instance,

$$\lim_{t \to \infty} f(t) = \lim_{s \to 0} sF\,(s). \tag{16}$$

Applying this theorem,

$$\lim_{t \to \infty} \theta\;(t) = \lim_{s \to 0} P\;(s) \tag{17}$$

If the open-cycle controller is to have no static error, the final value of $\theta_{o1}(t)$ must equal the applied displacement. Therefore,

$$\lim_{s \to 0} P(s) = 1\;. \tag{18}$$

Stated in a slightly different way, Equation (18) amounts to prescribing that the zero-frequency value of the transfer-function of the open-cycle controller must equal unity (or some other value invariant with frequency in case the controller amplifies the input motion.) This requirement, of course, is obvious without such a detailed calculation, but the calculation was carried through to introduce certain analysis principles.

The significance of the condition for zero static error in the open-cycle controller is apparent at once. The quantity $\lim_{s \to 0} P(s)$ must be invariant with changes of load, ageing, temperature effects and the like if high accuracy is to be maintained, and therefore, the open-cycle controller essentially must be a calibrated system. In general, such requirements are extremely difficult to satisfy, and consequently, open-cycle control systems are seldom employed in applications requiring high accuracy.

The condition for static accuracy in a closed-cycle control system, or servomechanism, is calculated by exactly the same method just employed for the open-cycle control system.

Figure 3 is a block diagram of such a system. A rather complex system is illustrated for the sake of generality and is divided into two components according to the dotted line enclosures. If the transfer-function (in the forward direction) of the upper enclosure is P(s) and that of the lower enclosure (in the reverse direction) is Q(s), the defining system equation is

$$\theta_o(s) = P(s)\left[\theta_i(s) + Q(s)\;\theta_o(s)\right]\;. \tag{19}$$

Equation (19) can be written

$$\theta_o(s) = \frac{P(s)\theta_i(s)}{1 - P(s)\;Q\;(s)} \tag{20}$$

---

* See reference 9, p. 265.

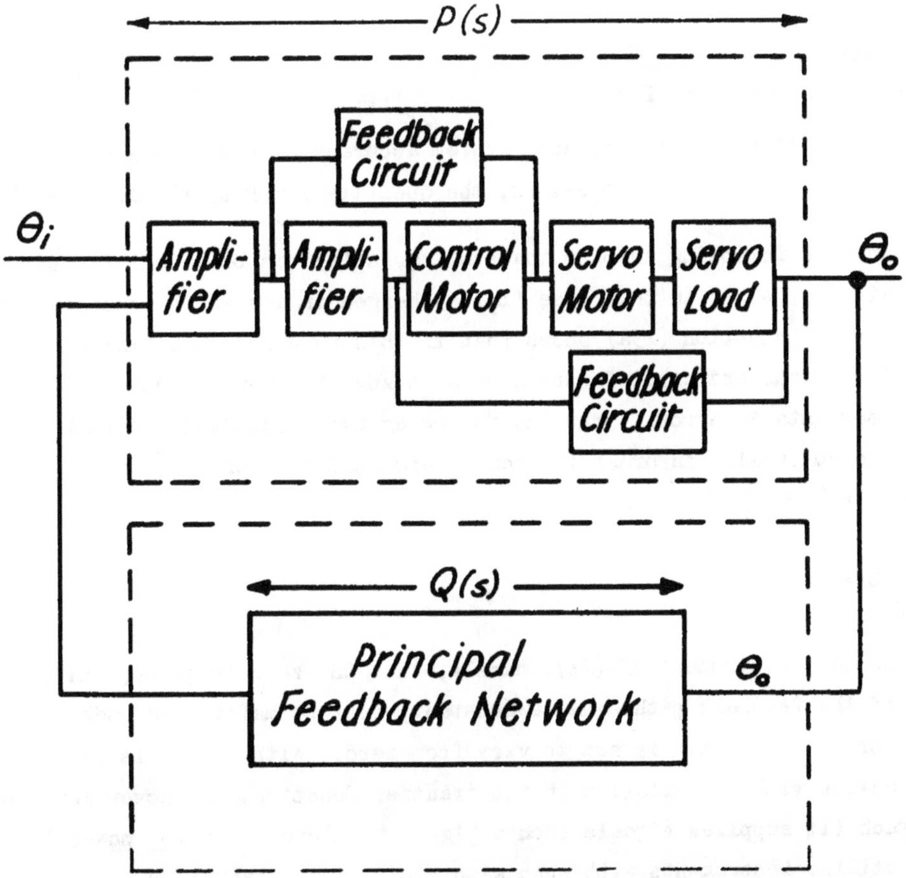

FIGURE 3

If $\theta_i(t)$ is a unit function, as illustrated by Figure 2, the transform of the servo output is

$$\theta_{ol}(s) = \frac{P(s)}{s\left[1 - P(s)\,Q\,(s)\right]} \quad . \tag{21}$$

The final value of $\theta_{ol}(t)$ is

$$\lim_{t \to \infty} \theta_{ol}(t) = \lim_{s \to 0} \frac{P(s)}{1 - P(s)\,Q\,(s)} \tag{22}$$

If the servomechanism is to have no static error, the following restriction must be maintained.

$$\lim_{s \to 0} \frac{P(s)}{1 - P(s)\,Q\,(s)} = 1 \tag{23}$$

Solving,

$$\lim_{s\to 0} Q(s) = \lim_{s\to 0} \frac{1 - P(s)}{P(s)} \tag{23a}$$

In order to minimize the effects upon system accuracy caused by changes in servo load and changes in the constants of the system, the quantity $\lim_{s\to 0} P(s)$ should be as large as possible. It is shown in a following chapter that this quantity can be made infinite, which is, of course, highly desirable, since in that case finite changes in its value, caused by external factors, have no effect. Equation (23a) shows that if this function is infinite, the value of the transfer-function of the principal feedback path at zero frequency must equal negative unity in order to attain zero static error. Thus the following two conditions should be maintained in the closed-cycle controller in order to secure zero static error:

$$\left.\begin{aligned} \lim_{s\to 0} P(s) &= \infty \\ \lim_{s\to 0} Q(s) &= -1 \end{aligned}\right\} \tag{24}$$

The second requirement of (24), namely, that the zero-frequency value of the transfer-function of the feedback path equal a constant, must be maintained under all conditions if the static error of the system is not to vary from zero. Although it is impossible to secure a physical device with no variation in its transfer-function, the power level of the feedback path is such (it supplies signals from a high power level to a low power level) that it is possible to utilize instruments with such electrical and mechanical properties that the transfer-function of the feedback path approximates the required value for long periods of time.

The simplest physical device that satisfies the second requirement of Equation (24) is a device whose transfer-function is constant and equal to negative unity not only at zero frequency but at all frequencies. The block diagram of such a system is illustrated by Figure 4. The feedback path is simply a mechanical link through a differential from the servo output to the servo input. There are several types of servomechanisms that employ this simple feedback means and many others that employ synchro data transmission systems that approximate this basic type of feedback link.

Because of their importance, the subsequent discussion in the paper is devoted entirely to servomechanisms whose feedback links are equal to, or approximate, negative unity at all frequencies.

General Response Characteristics of Servomechanisms

Figure 4 is the block diagram of the type of servomechanism that is dealt with exclusively in this paper. Fundamentally, the servomechanism comprises a servo motor and its load, a system for controlling the servo motor, and a feedback link. In Figure 4 the system for controlling the servo motor is divided into several components, one or several of which may be absent in a particular servomechanism, but all of which are included in the diagram for generality and for later reference. The symbols for the servo input and output as a function of time are

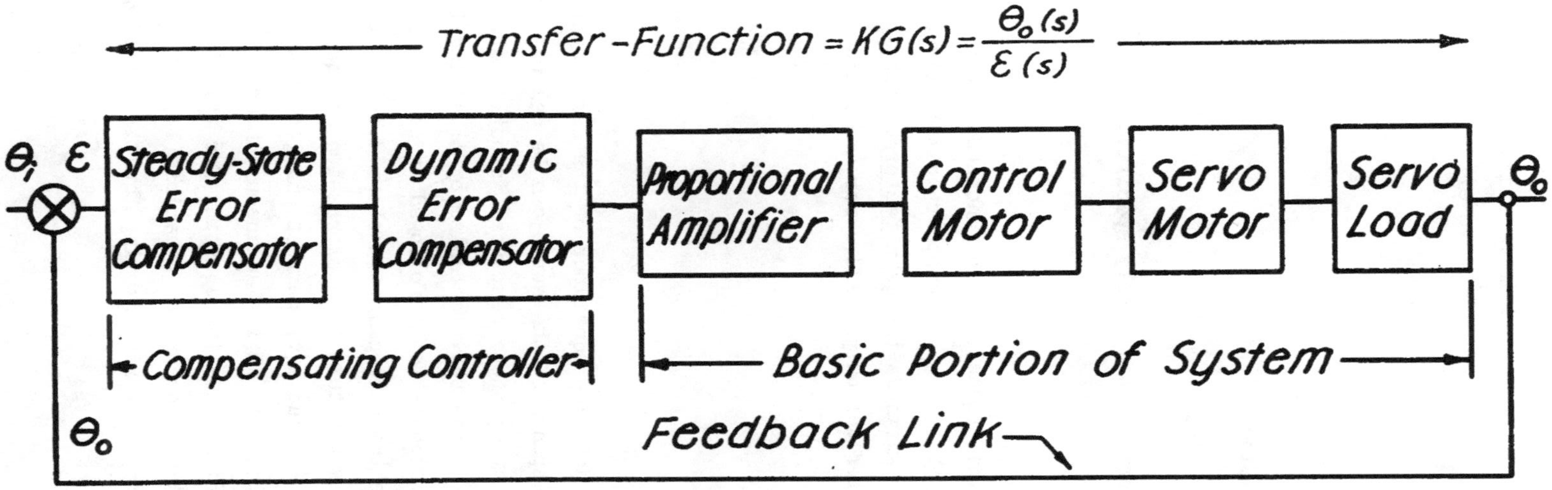

Transfer-Function = KG(s) = θo(s) / ε(s)
θi
ε
Steady-State Error Compensator
Dynamic Error Compensator
Proportional Amplifier
Control Motor
Servo Motor
Servo Load
θo
Compensating Controller
Basic Portion of System
θo
Feedback Link

FIGURE 4

$\theta_i(t)$ and $\theta_o(t)$, respectively, and the Laplacian transforms of these functions are defined by Equations (9) and (10). In addition, a third term is introduced which is the difference between the servo input and output and is known as the servo error. Its symbol is $\varepsilon(t)$, and its Laplacian transform is $E(s)$.

$$\varepsilon(t) = \theta_i(t) - \theta_o(t) \tag{25}$$

$$\mathcal{L}\left[\varepsilon(t)\right] = E(s) \tag{26}$$

The Laplacian equivalent of Equation (25) is

$$E(s) = \theta_i(s) - \theta_o(s). \tag{27}$$

The response of **the servomechanism is completely specified by prescribing the transfer-function** of the system.

$$KG(s) = \frac{\theta_o(s)}{E(s)} \tag{28}$$

The transfer-function $KG(s)$ is the product of two functions: one that is invariant with frequency and represented by the symbol "K", and another that is frequency dependent and is represented by $G(s)$. The term "K" is known as the <u>gain factor</u> or simply the <u>gain</u> of the system.

The Laplacian expressions for the output and error are found from equations (27) and (28).

$$E(s) = \frac{1}{1 + KG(s)} \theta_i(s) \tag{29}$$

$$\theta_o(s) = \frac{KG(s)}{1 + KG(s)} \theta_i(s) \tag{30}$$

Equations (29) and (30) are the defining equations of the servo system and both the transient response and the steady-state response of the servo can be calculated from these equations. The procedure involved in calculating these responses is outlined below.

Suppose it is desired to test the servomechanism by obtaining its response to some type of transient input. The general procedure to be followed in calculating either the servomechanism error or output is (1) determine the Laplacian transform $\theta_i(s)$ of the particular input function to be applied to the servo, (2) substitute this expression in Equations (29) and (30) and (3) obtain the inverse transforms $\theta_o(t)$ and $\varepsilon(t)$ corresponding to the expressions obtained for $\theta_o(s)$ and $E(s)$. A common test input is an instantaneous change of unit magnitude in the input displacement as illustrated in Figure 2. The equation for this type of input function is given by Equation (12) and its Laplacian transform by Equation (13). The Laplacian transforms of the output and error functions are found by substituting Equation (13) into Equations (29) and (30).

$$E_i(s) = \frac{1}{1 + KG(s)} \frac{1}{s} \tag{31}$$

$$\theta_{oi}(s) = \frac{KG(s)}{1 + KG(s)} \frac{1}{s} \tag{32}$$

The time responses $\varepsilon_1(t)$ and $\theta_{o1}(t)$ are determined by finding the inverse transforms of Equations (31) and (32). The process involved is outlined briefly for the case of the output function.

The determination of the inverse transforms of Equation (32) is simplified if the right-hand side of the equation is expanded as a partial fraction.[9] This is always possible for a particular physical system since for such a system Equation (32) will be a rational function.

$$\theta_{o1}(s) = \frac{A_0}{s} + \frac{A_1}{s + a_1} + \frac{A_2}{s + a_2} + \frac{A_3}{s + a_3} + \dots \tag{33}$$

The terms a_1 , a_2 , a_3 , etc., are the negatives of the roots of the equation

$$1 + KG(s) = 0 , \tag{34}$$

known as the characteristic equation of the system, and the terms A_0 , A_1 , A_2 , etc., are known as the residues of the function $\theta_{o1}(s)$ at its respective poles. The simple form of Equation (33) permits the inverse transform of $\theta_{o1}(s)$, which is $\theta_{o1}(t)$, the output of the servo as a function of time, to be found at once. It is given by Equation (35).

$$\theta_{o1}(t) = A_0 + A_1e^{-a_1t} + A_2e^{-a_2t} + A_3e^{-a_3t} + \dots \tag{35}$$

Thus $\theta_{o1}(t)$ consists of a "steady-state" term A_0 and a series of transient terms $A_1e^{-a_1t}$, $A_3e^{-a_3t}$, etc. The transient terms approach zero as time increases (the servomechanism is stable) provided the exponents a_1 , a_2 , a_3 , etc., are either positive real numbers or complex numbers with positive real parts. The condition that must be met if the steady-state servo output is to be equal to the input has been considered on p. 12. If this condition exists, the steady-state term, A_0 , of Equation (35) is equal to unity, and the steady-state error consequently is equal to zero.

The character of the roots of Equation (34) (the exponents $-a_1$, $-a_2$, $-a_3$, etc.) and the residues A_1 , A_2 , A_3 etc., determine the manner with which the servo output $\theta_{o1}(t)$ approaches its final value. In the ideal case the residues A_1 , A_2 , A_3 , etc., would be zero, or the exponents a_1 , a_2 , etc., would be infinite, and the servo output would follow the input instantly. Such a system is physically unattainable and the best that can be done is to so design the system that the output approaches its final value sufficiently rapidly that reasonable application requirements are met. Typical servo response curves are illustrated in Figure 5. Curve A is the response of a system all the roots ($-a_1$, $-a_2$, $-a_3$, etc.) of which are real and unequal; Such a system is said to be "overdamped." If the roots of the characteristic equation (Equation (34)) are all real and equal, the system is said to be "critically damped" and the response if illustrated by Curve B of Figure 5. Complex roots give rise to responses such as are illustrated by curves C and D of Figure 5 and the system is said to be "underdamped." The amount the servo output "overshoots" its final value is an inverse function of the magnitude of the real part of the complex roots. Thus curve D is the response of a system the complex roots of which have smaller real parts than those of the system the response of which is illustrated by curve C.

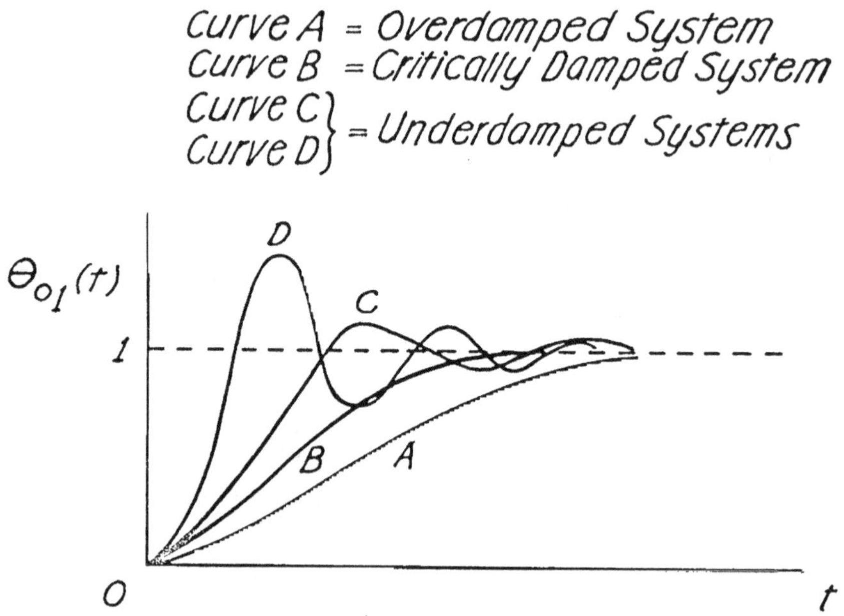

FIGURE 5

Certain requirements must be met by the roots $-a_1$, $-a_2$, $-a_3$, etc., in order that the servo performance be satisfactory. Those roots that are real must be sufficiently large that the factor e^{-at_0} is negligible at the end of a time interval t_0 determined by the servo application. Those exponentials that are complex must have a sufficiently large real part (damping constant) that they, too, are negligible after the prescribed time interval, t_0 . A "fast" servo is characterized by large real roots and complex roots whose real parts are large. The residues A_1 , A_2 , etc., also affect the speed of the system but not so greatly as the roots of the characteristic equation. Figure 5 shows that in general fastest servo response is obtained provided the system is so adjusted that it is slightly underdamped. Brown[6] has considered the transient response of servomechanisms in some detail and has given adjustment criteria for simpler systems that result in rapid servo response.

The steady-state response of the servomechanism is also of value as a guide to proper system adjustment. The relative magnitude and phase of both the servo error and the servo output are found from Equations (29) and (30) by replacing the complex variable, s, by the frequency, jω. This process has been explained on p. 8.

$$\frac{E}{\theta_i}(j\omega) = \frac{1}{1 + KG(j\omega)} \tag{36}$$

$$\frac{\theta_o}{\theta_i}(j\omega) = \frac{KG(j\omega)}{1 + KG(j\omega)} \tag{37}$$

Valuable information is gained by studying the phase and magnitude of both Equations (36) and (37) but at this point Equation (37) alone is considered. The functions to be examined are (1) the magnitude ratio, $\left|\frac{\theta_o}{\theta_i}(j\omega)\right|$, and (2) the phase difference, arc $\left[\theta_o(j\omega)\right]$ - arc $\left[\theta_i(j\omega)\right]$, (where "arc" signifies "the angle of") of the servo output and input. Similar functions exist for the servomechanism error.

An ideal servo is characterized by amplitude and phase functions illustrated in Figure 6, in which the amplitude function is unity, and the phase function is zero for all frequencies.

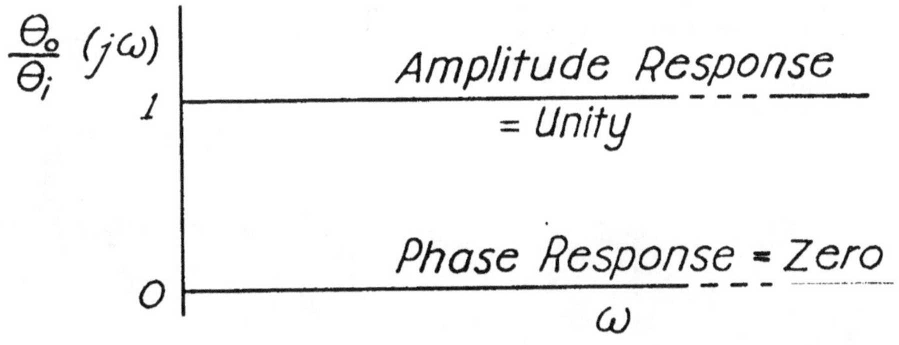

FIGURE 6

Such a system cannot be physically obtained; however, every operative servomechanism will have amplitude and phase functions which approximate the ideal over a limited frequency range. The problem is to make the approximation good over a sufficiently wide frequency range that the requirements of the application are met for which the servo is intended.

The correlation of the amplitude and phase response of a servo with application requirements such as speed of response and accuracy are dealt with in other parts of this paper. This correlation is simplified by relating the sinusoidal response to the transient response. The amplitude and phase response of a physical servo are illustrated in Figure 7. Peaks in the amplitude response generally indicate the presence of complex roots of the characteristic equation whose complex parts correspond to the frequencies at which these peaks occur. The height of the amplitude response peak is a function of the real part of the complex root. If the real roots and the complex roots of the characteristic equation are to be large and have large real parts respectively, the peaks in the amplitude response function must be limited in magnitude and occur at large frequencies. It has been found by experience that the real part of the root is sufficiently large if the peaks in the amplitude response are limited to approximately one and one-third. With this restriction upon the amplitude peak, the "damping ratio" of the root, as defined by Brown[6], is approximately 0.5 to 0.8, and the imaginary part of the root is equal to the angular frequency at which the amplitude peak occurs within about twenty per cent. This correlation permits a rough calculation of the speed of response of the system to be made.

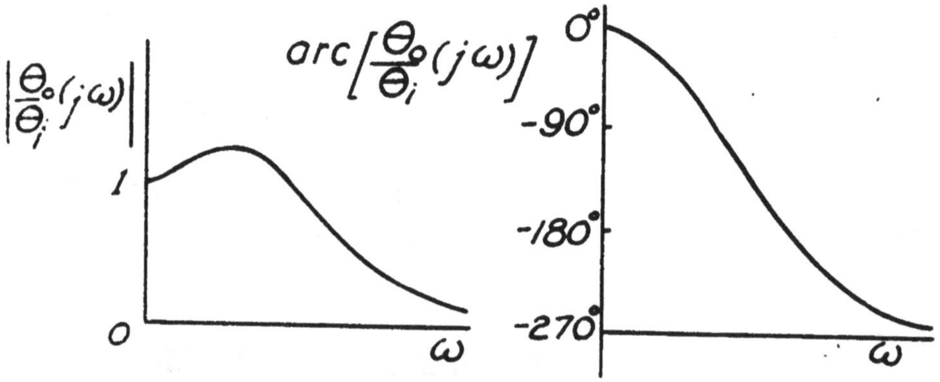

FIGURE 7

CHAPTER II

MINIMUM INTEGRAL-SQUARED-ERROR SERVOMECHANISM RESPONSE CRITERION

In general it is difficult to set up and apply precise mathematical criteria for optimum servomechanism performance because of the individual nature of each servo application and the complexity of exact criteria. However, the optimum set of adjustments for a servo designed for any application will, in general, fall within a range that is fixed and can be predicted with fair accuracy. While the accuracy and speed of response requirements are prescribed entirely by the individual servo application, the damping constants of the roots of the characteristic equation should all lie within a well-defined range for most servo applications. The requirements of individual servo applications may best be met by particular values of damping constants, but, in general, these particular values are not widely separated. Therefore, ideal adjustments for one servo application are close to the ideal for another application, and a servomechanism response criterion developed for a particular servo application is useful as a guide to proper servo adjustment in other problems.

A less complex mathematical criterion to set up and apply is based upon the time-integral of the square of the servo error produced by a particular servo input disturbance. For example, if in Figure 8 curve A is the time plot of an instantaneous or step-displacement in the servo input, $\theta_i(t)$, and curve B is a possible time-plot of $\varepsilon(t)$, the resulting servo error, the function $[\varepsilon(t)]^2$ is curve C of Figure 8, and the total area bounded by curve C and the time axis (the shaded area of Figure 8) is a measure of the speed of response of the system. The

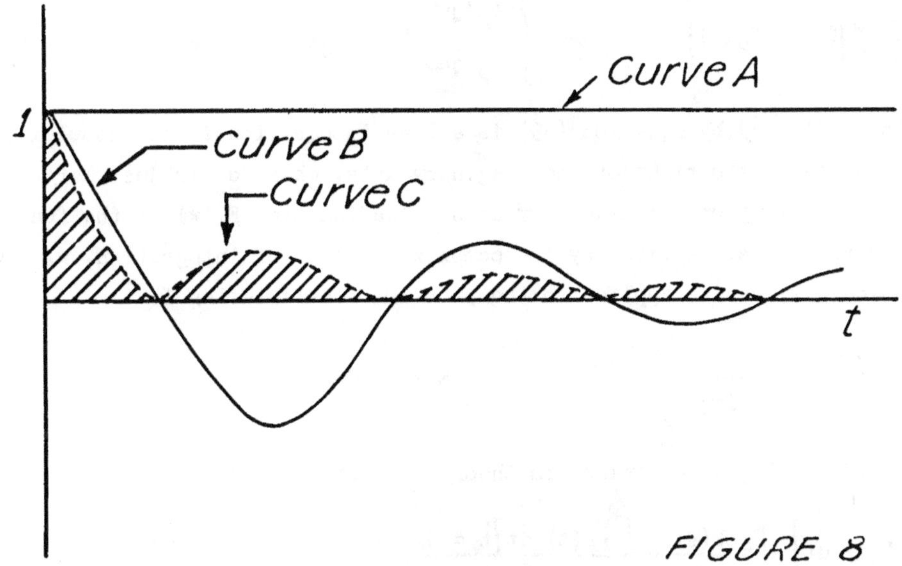

FIGURE 8

value of this area, denoted by the symbol, I, is expressed mathematically by Equation (38).

$$I = \int_0^{\infty} [\varepsilon(t)]^2 \, dt \qquad (38)$$

This criterion is applied by so determining the adjustable parameters of the servomechanism that the quantity, I, is a minimum.

The first step in applying the criterion is to express the quantity, I, in terms of the servo constants. The necessary calculations can be performed in two ways: (1) by expressing the Laplacian transform of the error in terms of the system constants, determining the corresponding equation for the servo-error as a function of time, and finally performing the time integration of the square of the error as indicated in Equation (38); (2) by calculating the function, I, by integration in the complex plane. The first of these two methods is straightforward; the second is less obvious and its theory is therefore developed below.

Let $\varepsilon(t)$ equal the servo error resulting from an arbitrary input function, $\theta_1(t)$. Let $E(s)$ equal the Laplacian transform of the time function $\varepsilon(t)$ and $\Theta_1(s)$ equal the Laplacian transform of the function $\theta_1(t)$. The relation between $E(s)$ and $\Theta_1(s)$ is provided by Equation (29):

$$E(s) = \frac{1}{1 + KG(s)} \Theta_1(s) \tag{29}$$

Since $KG(s)$, the transfer-function of the system, can be expressed in terms of the system parameters, and since the inverse transform, $\varepsilon(t)$, of $E(s)$ can be found, a relation between the system parameters and the time response, $\varepsilon(t)$ is derivable.

The following theorem in Laplacian transform theory is directly applicable to the calculation of I.*

If $F_1(s) = \mathcal{L}\left[f_1(t)\right]$

and $F_2(s) = \mathcal{L}\left[f_2(t)\right]$

Then $$\mathcal{L}\left[f_1(t)\, f_2(t)\right] = \frac{1}{2\pi j} \int_{c_2-j\infty}^{c_2+j\infty} F_1(w)F_2(s-w)\,dw \tag{39}$$

The operation indicated by Equation (39) is a line integration in the complex plane along a line displaced c_2 units to the right of the imaginary axis, where c_2 is just large enough to avoid any singularities lying on the imaginary axis. The function $F_1(w)$ is the function $F_1(s)$, with the complex variable, s, replaced by the complex variable of integration, w.

Equation (39) can be applied to the problem of calculating I.

$$\mathcal{L}\left[\left(\varepsilon(t)\right)^2\right] = \frac{1}{2\pi j} \int_{c_2-j\infty}^{c_2+j\infty} E(w)\; E(s-w)\; dw \tag{40}$$

A second theorem in Laplacian transform theory is useful here:**

$$\mathcal{L}\left[\int f(t)\; dt\right] = \frac{F(s)}{s} + \frac{\left[\int f(t)\; dt\right]_{t\,=\,0+}}{s} \tag{41}$$

---

* See reference 9, p. 275-278 ** See reference 9, p. 129

Making use of Equation (41) and realizing that the second term on the right hand side of Equation (41) must be zero,

$$\mathcal{L}\left\{\int[\varepsilon(t)]^2 dt\right\} = \frac{1}{s}\,\frac{1}{2\pi j}\int_{c_2-j\infty}^{c_2+j\infty} E(w)E(s-w)dw \tag{42}$$

For the last step in the calculation, use can be made of the initial and final value theorems of Laplacian transform theory.*

$$\lim_{t\to 0} f(t) = \lim_{s\to\infty} sF(s) \tag{43}$$

$$\lim_{t\to\infty} f(t) = \lim_{s\to 0} sF(s) \tag{16}$$

Then

$$\int_0^\infty [\varepsilon(t)]^2 dt = \lim_{s\to 0} s\mathcal{L}\left\{\int[\varepsilon(t)]^2 dt\right\} - \lim_{s\to\infty} s\mathcal{L}\left\{\int[\varepsilon(t)]^2 dt\right\} \tag{44}$$

Finally, since the value of $\int[\varepsilon(t)]^2 dt$ at $t = 0$, is zero, it follows that

$$\lim_{s\to\infty} s\mathcal{L}\left\{\int[\varepsilon(t)]^2 dt\right\} = 0.$$

Therefore,

$$I = \lim_{s\to 0} \frac{1}{2\pi j}\int_{c_2-j\infty}^{c_2+j\infty} E(w)\,E(s-w)\,dw \tag{45}$$

By complex variable theory,**

$$\int_{c_2-j\infty}^{c_2+j\infty} E(w)\,E(s-w)\,dw = 2\pi j\sum R \tag{46}$$

in which $\sum R$ represents the sum of the residues at the poles of the function $E(w)\,E(s-w)$ in the left half of the complex w plane. Thus Equation (45) becomes

$$I = \lim_{s\to 0}\sum R \tag{47}$$

<u>Example:</u> <u>Type I Servo</u>

The procedure that is followed in calculating the area under the error-squared curve is clarified if an example is carried through. A simple system that is often employed to demonstrate design principles comprises a servo motor whose load consists of moment of inertia, J_L, and viscous damping, f_L, and whose torque is directly proportional to the error. Such a system, termed a type I servo by Brown[6] is illustrated by Figure 9. This servo is discussed in

---

* See reference 9, p. 265 and 267.

** See reference 26, p. 113.

more detail in the discussion beginning with p. 89. At this point it will be employed as an example to illustrate the calculation and application of the criterion just developed.

TYPE I SERVO

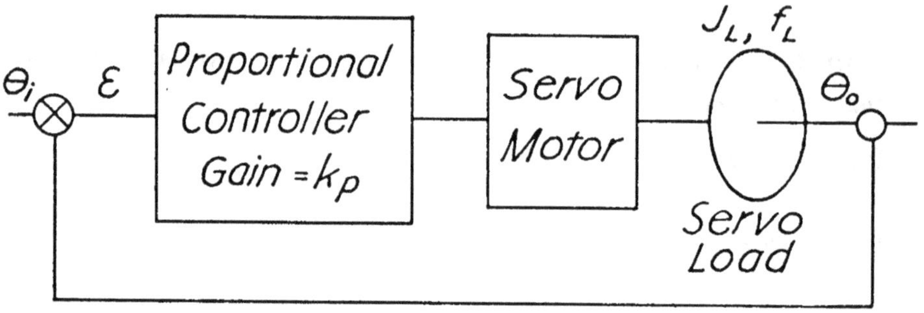

FIGURE 9

The Laplacian expression for the error of this system is given by Equation (48).

$$E(s) = \frac{(\mathcal{T}_L s + 1)\, s}{\mathcal{T}_L\left(s^2 + \frac{s}{\mathcal{T}_L} + \frac{k_p}{\mathcal{T}_L f_L}\right)}\, \theta_i(s) \tag{48}$$

In the above equation,

k_p = torque - error constant of the system (torque developed by the servo motor per unit of error.)

J_L = moment of inertia of the load and rotating parts of the motor.

f_L = viscous damping acting upon the motor output.

$$\mathcal{T}_L = \frac{J_L}{f_L}$$

If the input function, $\theta_i(t)$, is a unit step function, its transform, $\theta_i(s)$, is equal to $\frac{1}{s}$. The area, I, under the squared-error curve may be calculated by Equation (47) and is found to be

$$I = \lim_{s\to 0} \sum R \left\{ \frac{\mathcal{T}_L w + 1}{\mathcal{T}_L\left(w^2 + \frac{w}{\mathcal{T}_L} + \frac{k_p}{\mathcal{T}_L f_L}\right)} \cdot \frac{\mathcal{T}_L(s-w) + 1}{\mathcal{T}_L\left[(s-w)^2 + \frac{s-w}{\mathcal{T}_L} + \frac{k_p}{\mathcal{T}_L f_L}\right]} \right\} \tag{49}$$

in which $\sum R\{\ \}$ signifies "the sum of the residues in the left half plane." The residues occur at the roots of the characteristic equation.

$$w^2 + \frac{w}{\mathcal{T}_L} + \frac{k_p}{\mathcal{T}_L f_L} = 0 \tag{50}$$

These roots are

$$w = -\frac{1}{2\tau_L}\left[(1 \pm \sqrt{1 - \frac{4k_p\tau_L}{f_L}})\right] \qquad (51)$$

Let

$$\frac{k_p\tau_L}{f_L} = b \qquad (52)$$

The residues are evaluated from Equation (53), derived from Equation (49):

$$I = \lim_{s\to 0}\left\{\frac{1}{\tau_L^2}\;\frac{1 + \tau_L w}{w + \frac{1}{2\tau_L}(1-\sqrt{1-4b})}\;\frac{\tau_L(s-w) + 1}{(s-w)^2 + (s-w) + \frac{k_p}{\tau_L f_L}}\right\}_{w = -\frac{1}{2\tau_L}(1+\sqrt{1-4b})}$$

$$+ \lim_{s\to 0}\left\{\frac{1}{\tau_L^2}\;\frac{1 + \tau_L w}{w + \frac{1}{2\tau_L}(1+\sqrt{1-4b})}\;\frac{\tau_L(s-w) + 1}{(s-w)^2 + \frac{(s-w)}{\tau_L} + \frac{k_p}{\tau_L f_L}}\right\}_{w = -\frac{1}{2\tau_L}(1-\sqrt{1-4b})} \qquad (53)$$

When Equation (53) is evaluated, the result may be simplified to the following compact form:

$$I = \frac{\tau_L}{2}\,\frac{(1 + b)}{b} \qquad (54)$$

Equation (48) is frequently written in the following form:

$$E(s) = \frac{s(s + 2\zeta\omega_0)}{s^2 + 2\zeta\omega_0 s + \omega_0^2}\;\theta_i(s) \qquad (55)$$

In the above equation

$$2\zeta\omega_0 = \frac{f_L}{J_L}\,, \qquad (56)$$

$$\frac{k_p}{J_L} = \omega_0^2\;. \qquad (57)$$

The time function, $\varepsilon(t)$, is then studied for various values of ζ, termed the damping ratio, while the natural frequency, ω_0, is held invariant. If Equation (54) is rewritten in terms of the symbols ω_0 and ζ, Equation (58) is obtained:

$$I = \frac{1}{\omega_0}\,\frac{1 + 4\zeta^2}{4\zeta} \qquad (58)$$

It is, of course, desirable to so adjust the system constants that I, the area under the squared-error curve, is a minimum. Equation (58) shows that the value of I varies inversely with the natural frequency, ω_0, provided the damping ratio, ζ, is held constant. If ζ is varied and ω_0 held constant, a minimum value of I occurs when the damping ratio is equal to 0.5. The resulting value of I is given by Equation (59).

$$I_m = \frac{1}{\omega_0} \qquad (59)$$

Thus the criterion of minimum integral-squared-error yields the fact that the optimum value of ζ is 0.5 provided the natural frequency, ω_0 , is constant.

However, the ratio, ω_0 , is not a convenient quantity to hold invariant, since it is proportional to the torque-error constant, k_p, the only system constant easily adjusted. The ratio $\tau_L = \frac{J_L}{f_L}$ is more logically held constant since it depends only upon the load on the servo motor output and is generally fixed by externally imposed restrictions. A more correct consideration of the problem, therefore, is to so adjust the constant, k_p, that a minimum value of I occurs for a given value of τ_L. Equation (54) supplies the answer to this problem. Since b is directly proportional to k_p, the correct value of k_p is calculated from the value of b that results in a minimum value of I. A plot of I as a function of b is given in Figure 10.

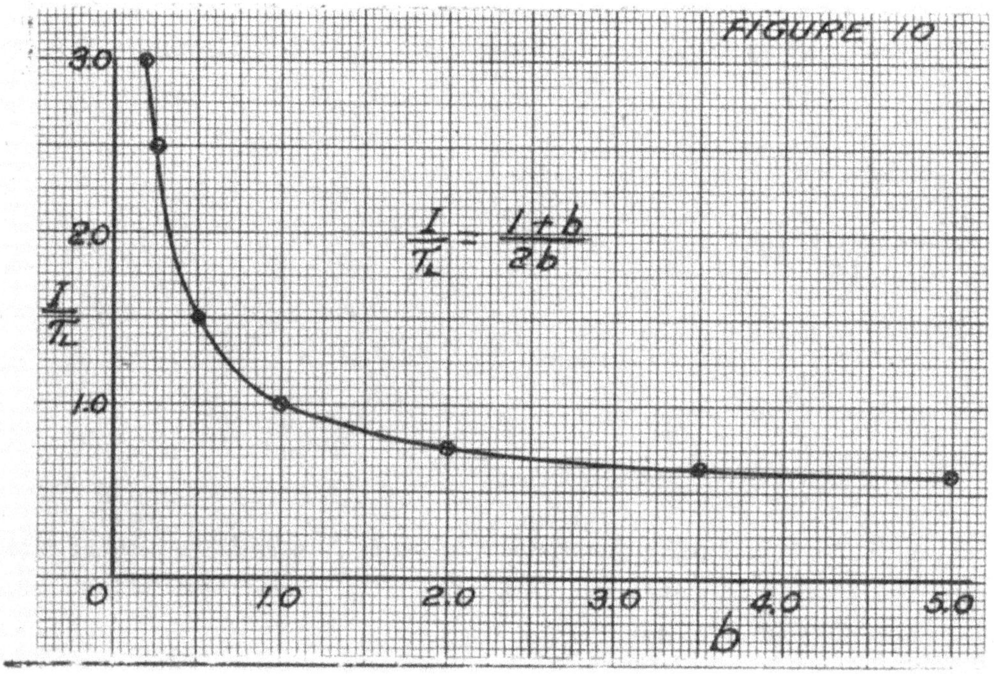

The significant fact is that <u>I is a constantly decreasing function with increasing b</u>. Therefore, the conclusion follows that the performance of the system is optimum if the torque-error constant, k_p, is infinite. Obviously, such a conclusion is erroneous for any physical system. The reason such a result appears is that the simple system of Figure 9 is stable for all values of k_p, although very large values of k_p produce a very low damping ratio. The natural frequency, ω_0, increases as k_p is increased, however, and at such a rate as to overcome the result of the decreasing damping ratio and to result in the variation in I depicted by Figure 10. Actually, no physical system exists that is stable for all values of gain and there is always a finite optimum value of gain for every physical system. The fact that I decreases constantly with increasing values of gain is an indication that for certain purposes the type I servo is an invalid approximation for a physical system. This point is discussed again in Chapter VI.

Example: Third-Order Servomechanism

The response of a physical servo system is much more accurately represented by an expression for E(s) or $\theta_o(s)$ of the third order than of the second order. A system whose characteristic equation is of the third order will always become unstable if the gain of the system is increased indefinitely, and therefore, the fallacy of a constantly decreasing integral I with increasing gain will not be encountered. It is of interest, therefore, to determine the conditions under which the integral, I, is a minimum for the case of the third-order system.

The transform of the error of a third-order system can generally be written

$$E(s) = \frac{s(s^2 + ps + q)}{s^3 + ps^2 + qs + r} \, \theta_i(s) \; . \tag{60}$$

Equation (60) is general for all servo systems that have no positional error (see p.36) and thus applies to practically all systems of interest. Because of the relation between the numerator and denominator of Equation (60), specification of the roots of the characteristic equation (the denominator of Equation (60) equated to zero) completely determines the system response. Let the three roots be known as λ_1, λ_2, and λ_3. Since the product of the three roots is r,

$$\lambda_1 \lambda_2 \lambda_3 = r \, , \tag{61}$$

the roots can be written as

$$\lambda_1 = a \underline{|\theta} \, , \tag{62}$$

$$\lambda_2 = a \underline{|-\theta} \, , \tag{63}$$

$$\lambda_3 = \frac{r}{a^2} \underline{|0^\circ} \tag{64}$$

In the above three equations, advantage is taken of the fact that optimum performance is secured if the roots are complex in character, and that in a third-order-system at least one root must be real. Therefore, a is the magnitude and $\pm\,\theta$ the angle of the complex roots.

The expression for I, the area under the squared-error curve, may be found by one of the two methods outlined previously. The actual calculation although straightforward is laborious. The work is not repeated here but the final result is given by Equation (65).

$$I = \frac{1}{D^2}\left\{\frac{A^2r^2}{a^3}\cos(2\phi - 3\theta) - \frac{2a^8}{r}\sin^2\theta - \frac{A^2r^2}{a^3\cos\theta} + \frac{8Aa^2r\sin\theta\sin(\phi - \theta - \Psi)}{B}\right\} \tag{65}$$

in which

$$A = \sqrt{a^2 + \frac{r^2}{a^4} - 2\,\frac{r}{a}\cos\theta} \tag{66}$$

$$B = \sqrt{a^2 + \frac{r^2}{a^4} + \frac{2r}{a} \cos\theta} \qquad (67)$$

$$D = 2a \sin\theta \left[\frac{2r}{a}\cos\theta - a^2 - \frac{r^2}{a^4}\right] \underline{|90^\circ} \qquad (68)$$

$$\phi = \tan^{-1} \frac{a^3 \sin\theta}{r - a^3 \cos\theta} \qquad (69)$$

$$\psi = \tan^{-1} \frac{a^3 \sin\theta}{r + a^3 \cos\theta} \qquad (70)$$

The complete determination of the mathematical minima of the quantity I is unsolved. It is possible to show that one minimum exists when θ equals zero and $a = \sqrt[3]{r}$; that is, the function I possesses a minimum when all the roots of the characteristic equation are real and equal Although other, and smaller, minima exist when the roots of the characteristic equation are complex, the mathematical conditions for their existence have not been found. The approximate determination of the minima of the function, I, has been carried out, however, by calculating and plotting the function for various assigned values of the magnitudes and angles of the roots of the characteristic equation. These results are summarized in the curves of figure 11. It is seen that the smallest of the four minima shown occurs when the magnitude of the complex root is equal to $(2r)^{1/3}$ and the angle is equal to eighty degrees. However, none of the minima show are widely different in value, and a rather wide latitude of choice of magnitude, a, and angle, θ, is permissible without appreciable loss in servo performance.

In conclusion, it should be emphasized that a servomechanism adjusted for optimum performance in accordance with the minimum integral squared-error criterion may have more overshoot, that is, be more under-damped, than is desirable for many servo applications. The reaso for this is that this criterion does not apply a time weighting factor. Thus, the error existing in the system immediately after the application of a step input is considered with equal weight as the error existing any time thereafter. This produces more over-shoot in a system adjusted to minimize the integral squared-error because such a system attempts to reduce the initial system error even at the expense of errors at a later time. Such an adjustment is superior for certain types of systems; others, however, are more properly adjusted if the system acts in such a way that the error after the application of the step input and prior to a time instant, t_0, is disregarded, and the error <u>after</u> the time interval t_0 is minimized. A system adjusted in the latter manner will have a different response from the one it would possess were it so adjusted that the integral I is minimum. Examples of applications in which one adjustment is superior to the other are obvious.

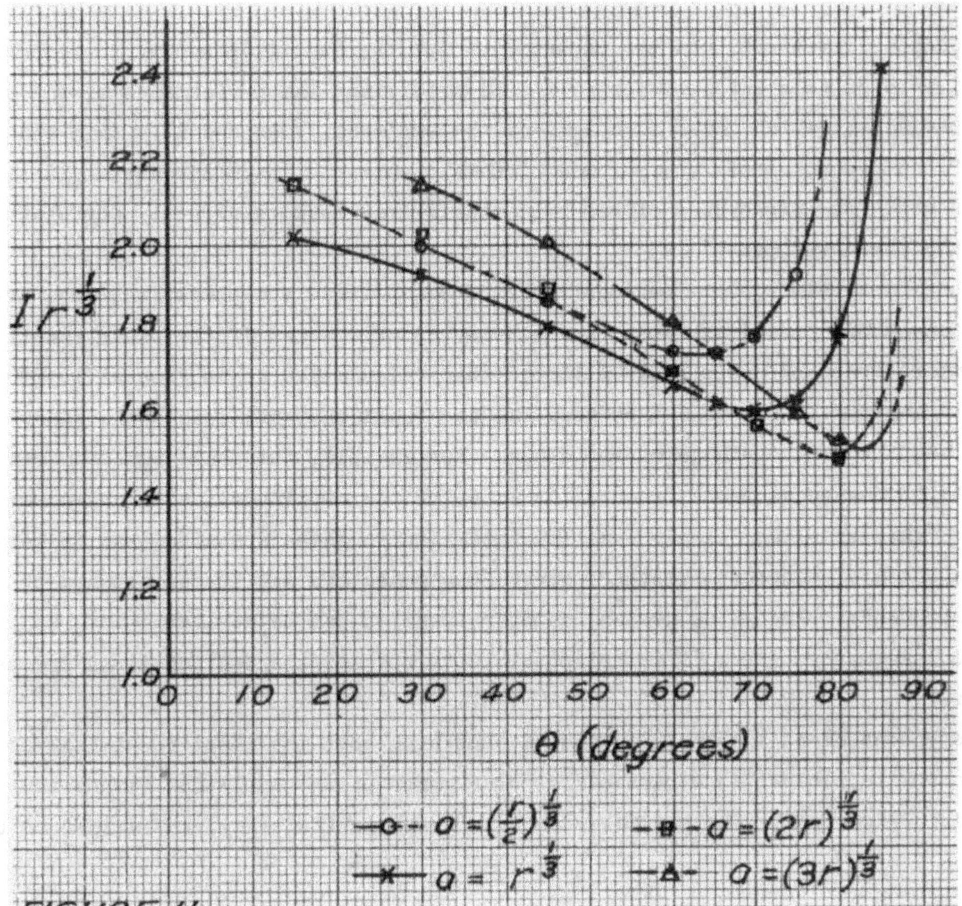

2.4
2.2
2.0
1.8
1.6
1.4
1.2
1.0
$Ir^{\frac{1}{3}}$
0
10
20
30
40
50
60
70
80
90
θ (degrees)
$a = (\frac{r}{2})^{\frac{1}{3}}$
$a = (2r)^{\frac{1}{3}}$
$a = r^{\frac{1}{3}}$
$a = (3r)^{\frac{1}{3}}$

FIGURE 11

CHAPTER III

FUNDAMENTAL PROPERTIES OF SERVOMECHANISM TRANSFER-FUNCTION-LOCI

In Chapter I was outlined the manner in which the amplitude and phase response of the servo could be used as a guide in design and in predicting the performance of the servomechanism. The manner in which these functions can be employed is expanded in this and subsequent chapters of the paper. It is the purpose of this chapter to introduce certain basic principles by means of which the properties of a servomechanism may be more directly correlated with its amplitude and phase response.

It has been pointed out previously that the amplitude and phase functions are not independent of one another, and that one function may be calculated if the other function is known over the entire frequency spectrum. Although this calculation is in general lengthy and arduous, certain basic relations exist between the two functions that enable the general form of one to be predicted from a knowledge of the other. For example, the phase function generally has a maximum rate of change at those frequencies at which the amplitude function has maximum or minimum values. This and other simple function interrelationships are discussed in some detail by Bode.[4] The interrelationship existing between the phase and amplitude response of a servo is of great importance. In system design, neither function can be neglected but both must be considered simultaneously, since it is not possible to specify the two functions independently of one another.

A very convenient and useful means of combining both the phase and amplitude functions into one composite picture is a parametric polar graph of the amplitude versus the phase as illustrated in Figure 12. The parameter in this graph is frequency, and values of the amplitude

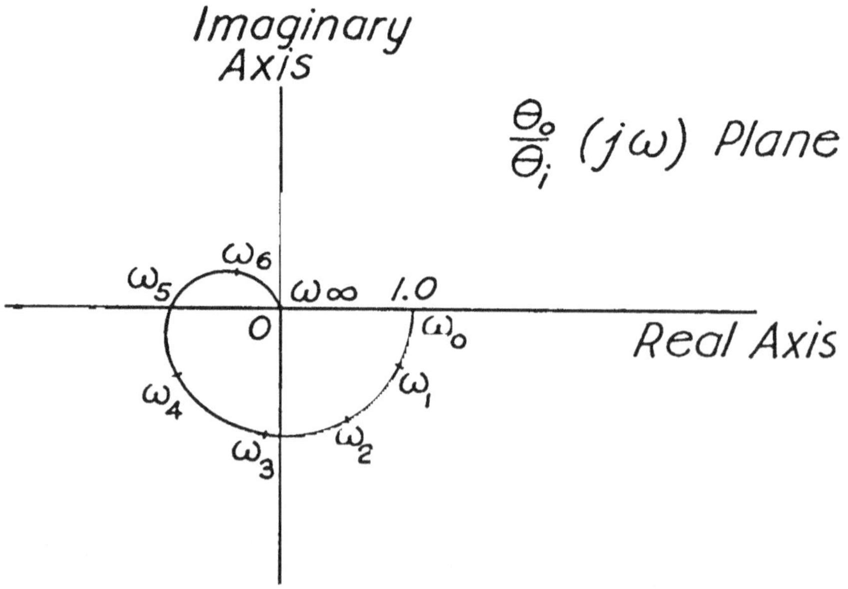

FIGURE 12

and phase functions for a particular frequency are given by the length and angle of the polar vector. This type of plot is used widely in the solution of electrical engineering problems with either frequency or some circuit constant as the parameter of the graph. The plots are very useful as aids in visualization and calculation of the performance of the circuit or device being investigated. This is particularly true if the geometric form of the plot is simple or is composed of combinations of simple forms, such as circles and straight lines. One curve contains the same information supplied by two curves if the amplitude and phase functions were plotted independently with the added advantage that restrictions upon and interrelations between phase and amplitude are emphasized by the use of a single curve. Therefore, a useful method of completely determining the response of a servomechanism is to specify the function $\frac{\theta_o}{\theta_i}(j\omega)$ by a parametric graph such as Figure 12. The function $\frac{E(j\omega)}{\theta_i(j\omega)}$ can also be plotted in this form.

If it is desired to analyse a servo's response and design compensation circuits for improving the servo performance, the function $\frac{\theta_o}{\theta_i}(j\omega)$ is awkward to work with directly. The reason for this is explained as follows. It has been shown that the performance of the servo is completely determined once the transfer-function, $KG(s)$, is known. This function relates the servo output $\theta_o(s)$, to the error, $E(s)$, the controller input. All the system design is applied to choosing, altering, and improving the servo-motor or servo-controller, the characteristics of which determine the transfer-function $KG(s)$. When the required form of the transfer-function is known, it is comparatively easy to synthesize the controller.

Design criteria expressed in terms of restrictions upon the transfer-function are most easily translated into physical design. However, the final decision as to the quality of the performance of a particular servo is made from a knowledge of the character of its output and error functions, and it is necessary, therefore, to translate restrictions upon the output and error functions into restrictions upon the transfer-function. The situation is summarized as follows: It is necessary to know the phase and magnitude of $\frac{\theta_o}{\theta_i}(j\omega)$ in order to decide if the servo is satisfactory: In synthesizing a servo it is easier to work with the transfer-function $KG(j\omega)$. How can performance criteria be translated into design criteria?

The alternative approach to the design problem using transient analysis is to synthesize the system in such a way that the roots of the characteristic equation, Equation (34), have satisfactory values. Graphical means of determining these roots when the order of the function is no higher than four mitigate the labor involved, but it is still a difficult procedure to translate information concerning the roots into actual design of the physical system. When the order of the characteristic equation is higher than four, as it is in almost any practical system, the work involved in obtaining the roots and translating that information into system design becomes very great.

It is therefore very desirable to correlate the characteristics of the transfer-function with those of the system response, in order that design methods may be based directly upon the properties of the transfer-function. It is the purpose of this chapter to relate the system characteristics to those of the transfer-function by means of an analysis of the locus of the transfer-function in the complex plane. Graphical calculation of pertinent characteristics are developed wherever their use simplifies the labor involved and aids in visualizing the problem.

Graphical Calculation of the Error and Output Functions

If a parametric frequency plot of the transfer-function, $KG(j\omega)$, is available, the magnitude and phase of the functions $\frac{\theta_o}{\theta_i}(j\omega)$ and $\frac{E}{\theta_i}(j\omega)$ may be found rather easily by graphical calculation The functions $\frac{\theta_o}{\theta_i}(j\omega)$ and $\frac{E}{\theta_i}(j\omega)$ are given by Equations (36) and (37).

$$\frac{E}{\theta_i}(j\omega) = \frac{1}{1 + KG(j\omega)} \tag{36}$$

$$\frac{\theta_o}{\theta_i}(j\omega) = \frac{KG(j\omega)}{1 + KG(j\omega)} \tag{37}$$

A plot of a typical transfer-function, $KG(j\omega)$, is illustrated by Figure 13. Suppose it is desired to calculate $\frac{E}{\theta_i}(j\omega)$ and $\frac{\theta_o}{\theta_i}(j\omega_c)$ for a particular frequency, ω_c. The transfer-function at ω_c, $KG(j\omega_c)$ is represented by the vector $\overline{oc}$, while the vector $\overline{ac}$, represents the term $\left[1 + KG(j\omega_c)\right]$, since the point "a" is located at $-1 + j0$. Therefore, the magnitude of $\frac{\theta_o}{\theta_i}(j\omega)$ is given by Equation (71).

$$\left| \frac{\theta_o}{\theta_i}(j\omega_c) \right| = \frac{|KG(j\omega_c)|}{|1 + KG(j\omega_c)|} = \frac{|\overline{oc}|}{|\overline{ac}|} \tag{71}$$

while the phase of $\frac{\theta_o}{\theta_i}(j\omega)$ is given by (72):

$$\text{arc}\left[\frac{\theta_o}{\theta_i}(j\omega_c)\right] = \text{arc (doc)} - \text{arc (oac)} = \text{arc (oca)} \tag{72}$$

The angle, arc (oca), is negative. Thus the magnitude of $\frac{\theta_o}{\theta_i}(j\omega_c)$ is equal to the angle between these two vectors.

Equations (71) and (72) permit ready visualization or calculation of the magnitude and

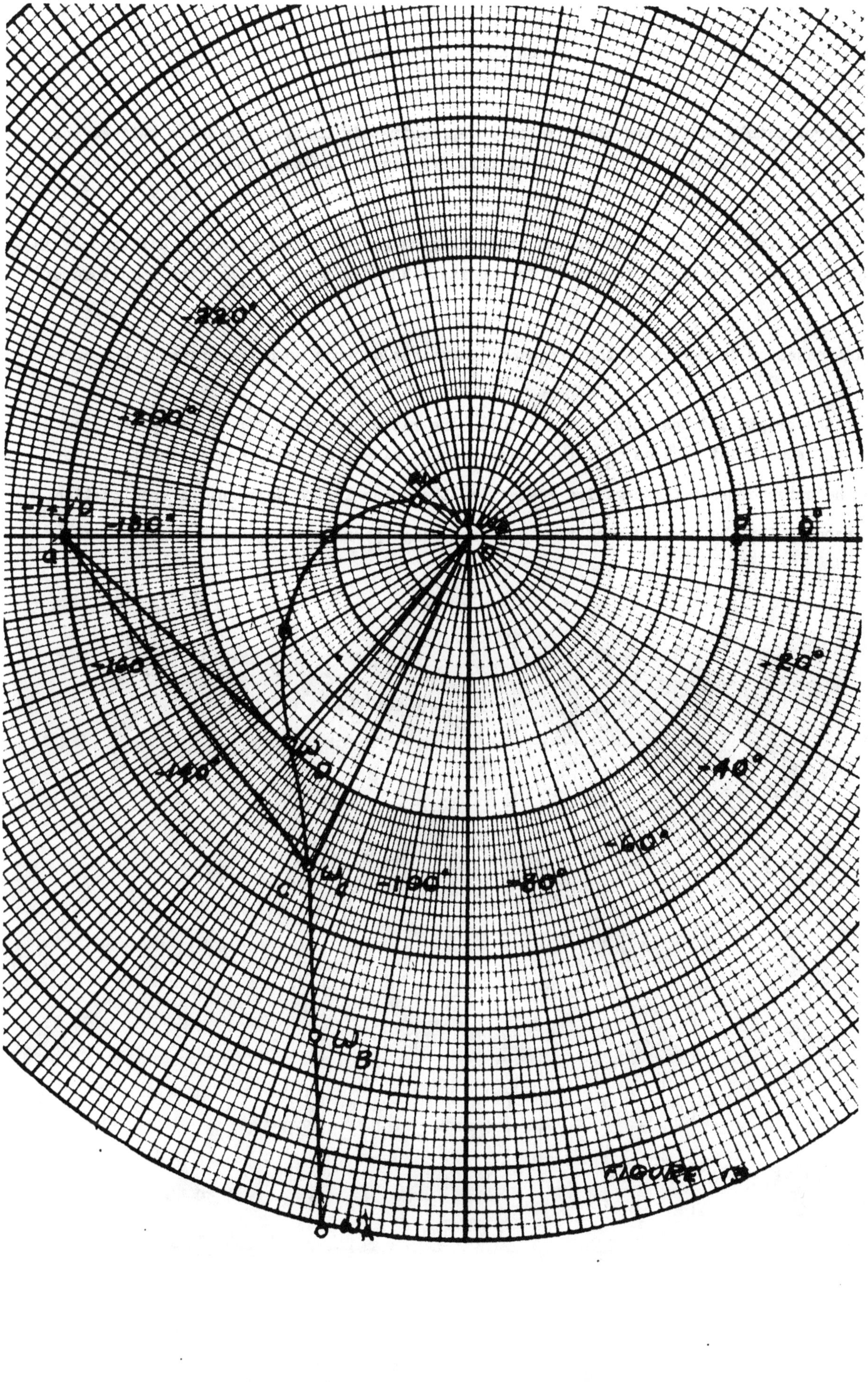
-220°
-200°
-180°
-160
-140°
-120°
-100°
-80°
-60°
-40°
-20°
0°
FIGURE 13

phase of $\frac{\theta_o}{\theta_i}$ $(j\omega)$. At small frequencies, such as ω_a (see figure 13), both vectors $\overline{oc}$ and $\overline{ac}$ are large and approximately equal, and their ratio is approximately unity. The angle between the two vectors, the phase of $\frac{\theta_o}{\theta_i}$ $(j\omega)$, is small at this frequency. As the frequency increases the angle between the two vectors increases and their lengths become smaller so that differences in their lengths cause the ratio $\frac{\overline{oc}}{\overline{ac}}$ to depart from unity. Whether the ratio $\frac{\overline{oc}}{\overline{ac}}$ (the magnitude of $\frac{\theta_o}{\theta_i}$ $(j\omega)$) increases or decreases as the frequency increases depends upon the shape of the curve relative to the origin and the point $(- 1 + j0)$. The ratio decreases for the case depicted by figure 13. At some point such as ω_d , the phase reaches ninety degrees, and at the frequency ω_e , when the curve crosses the negative real axis, the phase is 180 degrees. A continuation of this reasoning for the remainder of the frequency range permits the general shap

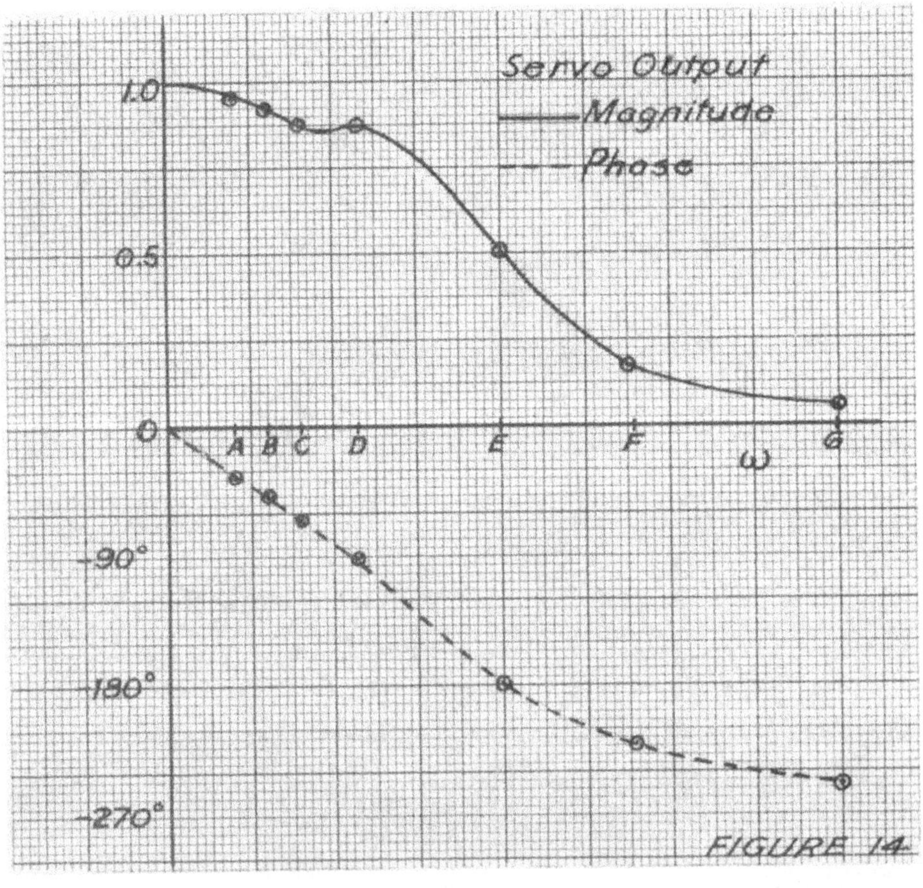

FIGURE 14

phase and magnitude of the servo output to be completely determined. These curves corresponding to figure 13 are sketched in figure 14. If necessary, they can be determined from figure 13 with accuracy and ease by the use of a protractor and divider and measuring the angles and lengths directly from the graph.

The function $\frac{E}{\theta_i}(j\omega)$, defined by Equation (37), can be calculated in a similar way. For a particular frequency, ω_c, the magnitude and phase of $\frac{E}{\theta_i}(j\omega)$ are given by Equations (73) and (74) respectively.

$$\frac{E}{\theta_i}(j\omega_c) = \frac{1}{1 + KG(j\omega_c)} = \frac{1}{\overline{ac}} \tag{73}$$

$$\text{arc}\ \frac{E}{\theta_i}(j\omega_c) = -\ \text{arc}\ (oac) \tag{74}$$

The shape of the phase and magnitude functions of the error are readily sketched. Referring to the transfer locus of figure 13, it is seen that the magnitude of the error at low frequencies is small, since it is inversely proportional to the length of the vector $\overline{ac}$. At large frequencies the magnitude of the error approaches unity. The phase of the error at low frequencies is slightly less than ninety degrees and leads the servo input. The phase of the error passes through zero at the frequency ω_e, and at large frequencies approaches zero asymptotically. The general forms of the error amplitude and phase functions are illustrated in figure 15. If accurate values are required, they may be obtained readily by scaling the graph of the transfer-function.

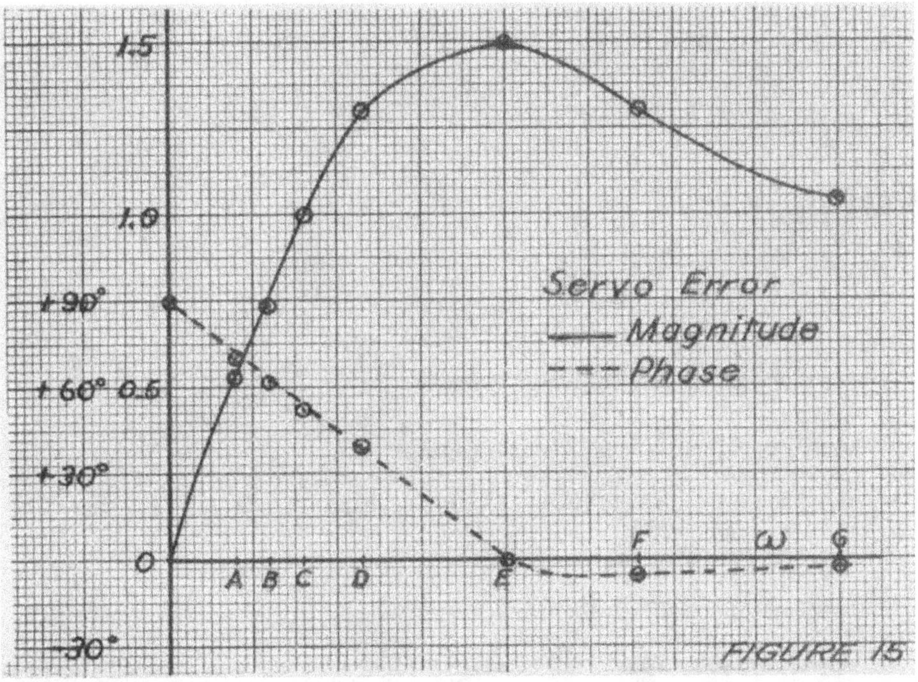

FIGURE 15

It has been shown how the magnitude and phase of both the servo error and the servo output can be readily calculated by graphical means from the transfer-locus of the servomechanism. In the next chapter are developed methods whereby the transfer-locus can be so adjusted that an optimum amplitude and phase response results.

Nyquist's Stability Criterion

The primary requirement that every servomechanism must satisfy is that of stability. Although a servo must be more than barely stable to be satisfactory, a stability criterion of one type or another is generally the first test applied to a newly proposed automatic control system. A simple stability criterion developed by Nyquist[21] primarily for feedback amplifiers is readily applied to the transfer locus of a servomechanism. The similarity between feedback amplifiers and servomechanisms has long been realized and, of course, the application of Nyquist's stability criterion is not original with this paper, but is included nevertheless for the sake of completeness.

The block diagram of a feedback amplifier drawn in conventional manner is shown in figure 16. The equations defining the system are

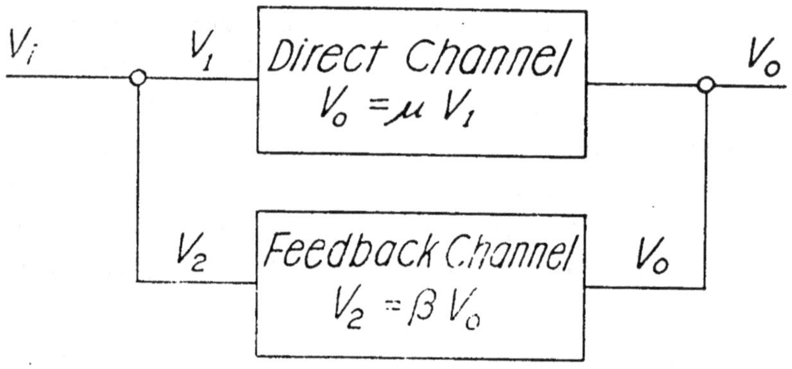

FIGURE 16

$$V_1(s) = V_i(s) + V_2(s) \qquad (75)$$

$$V_o(s) = \mu(s)V_1(s) \qquad (76)$$

$$V_2(s) = \beta(s)V_o(s) \qquad (77)$$

in which

$V_i(s)$ = transform of input voltage to system

$V_o(s)$ = transform of output voltage of system

$V_1(s)$ = transform of input voltage to direct channel

$\mu\ (s)$ = transfer-function of direct channel

$\beta\ (s)$ = transfer-function of feedback channel

A comparison of the servomechanism block diagram, figure 4, with the feedback amplifier diagram, figure 16, and the servomechanism equations (29) and (30), with the feedback amplifier equations (75) , (76) , and (77) , shows the two systems to be identical except for symbols provided that

$\beta(s)$ is set equal to negative unity. If the symbols are transferred as follows, complete

$$\begin{aligned} V_i(s) &= \Theta_i(s) \\ V_o(s) &= \Theta_o(s) \\ V_1(s) &= E(s) \\ \mu\ (s) &= KG(s) \\ \beta\ (s) &= -1 \end{aligned} \tag{78}$$

identity exists between the two systems. Thus it is evident that a servomechanism may be considered a feedback amplifier in which $\beta(s)$, the transfer-function of the feedback channel, is equal to minus one.

Nyquist's stability criterion, when applied to feedback amplifiers, is concerned with the shape and position of the locus of the function $\mu\beta(j\omega)$. If a parametric plot of $\mu\beta(j\omega)$ for both positive and negative frequencies is drawn, Nyquist has shown that the locus corresponds to a stable amplifier provided it does not enclose the point $(+1 + j0)$, and corresponds to an unstable amplifier should the locus enclose the point $(+ 1 + j0)$. This is illustrated in figure 17. The solid line curves are the portion of the loci plotted for positive frequencies and the

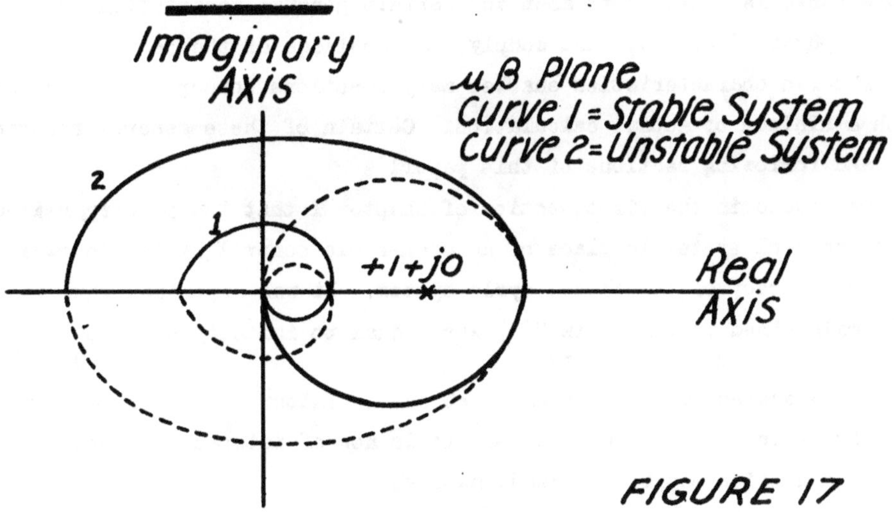

FIGURE 17

broken line curves the portion plotted for negative frequencies. Curve 1 does not enclose the positive unity point and therefore is the transfer-locus of a stable amplifier. Curve 2, however, does enclose the positive unity point and therefore is the transfer-locus of an unstable amplifier. The portions of the loci corresponding to negative frequencies need not actually be calculated, since the locus for a negative frequency is always the conjugate of the locus for the positive frequency of equal magnitude. Thus

$$\mu\beta(-j\omega_1) = \text{Conjugate of } \left[\mu\beta(+j\omega_1)\right]. \tag{79}$$

Nyquist also has shown that if there is doubt as to whether or not the locus encloses the unity point, it may be resolved as follows: Let a line be drawn from the point $(+ 1 + j0)$ to the locus of the function $\mu\beta(j\omega)$. As the frequency varies, let the end of the line on the curve trace out the locus while the end of the line at the unity point remains fixed. If the net angle through which the line swings as the frequency varies from $+\infty$ to $-\infty$ is zero, the system is stable; otherwise, the system is unstable.

Because of the similarity existing between the servomechanism and the feedback amplifier, the Nyquist stability criterion may be applied to either. Relations (78) show that if the criterion is to be applied to a servo system, the function $-KG(j\omega)$ is plotted, since μ corresponds to $+ KG(j\omega)$ and β is equal to minus one for a servomechanism. If the plot of $- KG(j\omega)$ encloses the point $(+1 + j0)$ the system is unstable; otherwise, the system is stable.

If the locus of $- KG(j\omega)$ encloses the point $(+ 1 + j0)$, the locus of the function $+ KG(j\omega)$ will enclose the point $(-1 + j0)$; and if the negative of the transfer-function locus fails to enclose the positive unity point, so will the transfer-function locus fail to enclose the negative unity point. Therefore, it is unnecessary to plot the negative of the transfer-function if the transfer-function locus itself is already available. Throughout this paper the transfer-function itself, $+ KG(j\omega)$, is plotted and the stability criterion is applied between its locus and the point $(-1 + j0)$.

Transfer-Locus of Zero-Displacement-Error Servo

The form of the transfer-locus of a servomechanism is controlled by the application requirements the servo is designed to meet and certain general restrictions with which the servomechanism, as a physical system, must comply. A knowledge of the form of the transfer-locus for particular servo characteristics enables many questions concerning servo design to be answered with a minimum of actual calculation. Certain of these general restrictions are developed in this and following sections of this paper.

It has been shown in the first section of Chapter I that the primary reason for employing a closed-cycle control system in place of an open-cycle control system is because of the inherently higher accuracy of the closed-cycle system. It was explained why highest system accuracy would be maintained provided $\lim_{s\to 0} P(s)$ were equal to infinity and $\lim_{s\to 0} Q(s)$ were equal to negative unity. A system was postulated by which the latter condition was met, and this paper is devoted entirely to that type of system. It is now of interest to investigate methods by which the first condition may be realized; namely,

$$\lim_{s\to 0} P(s) = \infty .$$

This requirement in terms of the symbol now employed to represent the transfer-function of the system of figure 4 is

$$\lim_{s\to 0} KG(s) = \infty . \tag{80}$$

Since K is a constant equal to the gain of the system and is independent of frequency, equation (24) can be true only if

$$\lim_{s\to 0} G(s) = \infty . \tag{81}$$

Mathematically this means that the function $G(s)$ must have a pole of order one or higher at $s=0$

The function G(s) is a rational function and will be of the general form given by Equation (82) ,

$$G(s) = \frac{(s + b_1)\ (s + b_2)\ (s + b_3)\ldots.}{s^n\ (s + c_1)\ (s + c_2)\ (s + c_3)\ldots.} \tag{82}$$

in which "n" is an integer. Equation (81) requires that the exponent "n" in Equation (82) be equal to or greater than unity if the servo is to have zero displacement-error.

As a function of frequency, the transfer-function is of the form given by Equation (83) provided the displacement error is zero:

$$KG(j\omega) = \frac{(j\omega + b_1)\ (j\omega + b_2)\ (j\omega + b_3)\ \ldots\ldots}{(j\omega)\ (j\omega + c_1)\ (j\omega + c_2)\ (j\omega + c_3)\ \ldots} \tag{83}$$

at the frequency, ω approaches zero, Equation (83) becomes

$$\underset{\omega \to 0}{KG(j\omega)} \cong \frac{b_1 b_2 b_3\ \ldots\ldots}{(j\omega)(c_1 c_2 c_3)\ \ldots\ldots} \tag{84}$$

Since both the products $b_1b_2b_3\ldots$ and $c_1c_2c_3\ldots$ are real, the locus of the transfer-function approaches infinity along the negative imaginary axis of the complex plane, as illustrated in figure 18. The necessary and sufficient condition, therefore, that a servomechanism possess

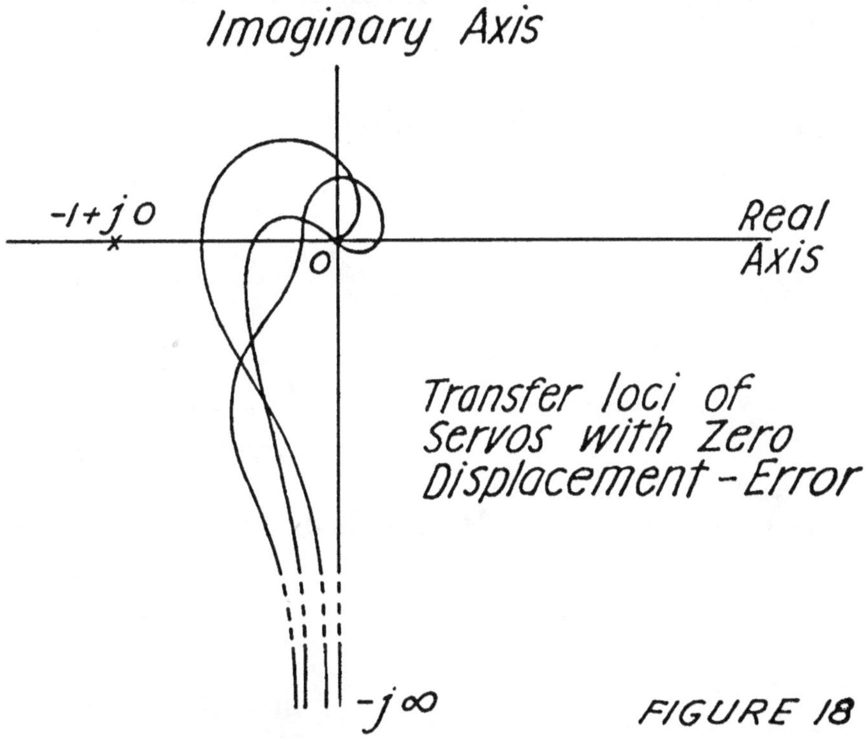

FIGURE 18

zero displacement error is that the locus of the transfer-function of the servo approach infinity along the negative imaginary axis as the frequency approaches zero.

A servomechanism possessing a transfer-function of the type given by Equation (83) is obtained by incorporating in cascade in the controller, or by using as a servo motor, a device that has infinite output displacement for a finite input displacement. For example, suppose the servo motor is a d-c motor with constant field excitation and loaded with inertia and viscous friction. Let the servo-controller be such that the motor armature current is directly proportional to the controller input. The output displacement of such a motor is measured in radians and the input displacement is measured in amperes. Since a given current displacement in the motor armature will cause the motor output to revolve continuously (infinite output displacement), such a system possesses the required form of transfer-locus to have zero displacement-error. However, if the motor load is partly Coulomb friction, the motor will not revolve unless sufficient armature current is applied to produce a torque large enough to overcome the Coulomb friction. Therefore, the servomechanism no longer possesses a transfer-function of the form given by Equation (83) and static error may exist in the system. The transfer-function of Equation (83) may also be obtained through the use of regenerative amplifiers in the servo-controller.

<u>Transfer-Locus of Zero-Velocity-Error Servo</u>

Many servomechanism applications require the servo to operate with little or no error when the servo input is a constant velocity. The form of the transfer-function if the servo error is to be zero under such input conditions may be found by a method similar to that employed in the previous section. If $\theta_i(t)$ is a suddenly applied velocity, as illustrated in Figure 19, its

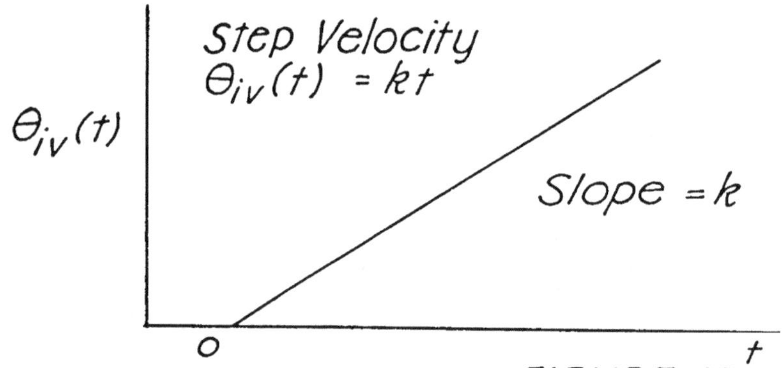

FIGURE 19

Laplacian transform is given by Equation (85) in which $\theta_{iv}(s)$ is the transform of $\theta_{iv}(t)$.

$$\left.\begin{aligned} &\theta_{iv}(s) = \frac{k}{s^2} \\ &\text{if} \\ &\theta_{iv}(t) = 0, \quad 0 > t > -\infty \\ &\theta_{iv}(t) = kt, \quad 0 < t < +\infty \end{aligned}\right\} \quad (85)$$

The Laplacian transform of the servo error is found by substituting the expression for $\theta_{iv}(s)$ into the general equation for servo error, Equation (29).

$$E_v(s) = \frac{1}{1 + KG(s)} \quad \frac{k}{s^2} \tag{86}$$

The final value of the error is the steady-state error and is found by applying Equation (16).

$$\varepsilon_v(\infty) = \lim_{t \to \infty} \varepsilon_v(t) = \lim_{s \to 0} \frac{1}{1 + KG(s)} \quad \frac{k}{s} \tag{87}$$

If the steady-state velocity-error is to equal zero,

$$\lim_{s \to 0} sG(s) = \infty \tag{88}$$

Mathematically this means that the function G(s) must have a pole of order two or higher at s=0. The exponent, n, in the general expression (82) for the transfer-function must be equal to, or greater than, two. The transfer-function, as a function of frequency, must be of the form given by Equation (89), therefore, if the servo is to have no velocity-error.

$$KG(j\omega) = \frac{(j\omega + b_1)\ (j\omega + b_2)\ (j\omega + b_3)\ldots\ldots\ldots\ldots}{(j\omega)^2\ (j\omega + c_1)\ (j\omega + c_2)\ (j\omega + c_3)\ \ldots\ldots} \tag{89}$$

As the frequency, ω, approaches zero, (89) becomes

$$\underset{\omega \to 0}{KG(j\omega)} \cong - \frac{b_1 b_2 b_3 \ldots\ldots}{(\omega^2)\ c_1 c_2 c_3 \ldots} \tag{90}$$

Since both products $b_1 b_2 b_3 \ldots$ and $c_1 c_2 c_3 \ldots$ are real, the locus of the transfer-function approaches infinity along the negative real axis, as illustrated in figure 20. The condition,

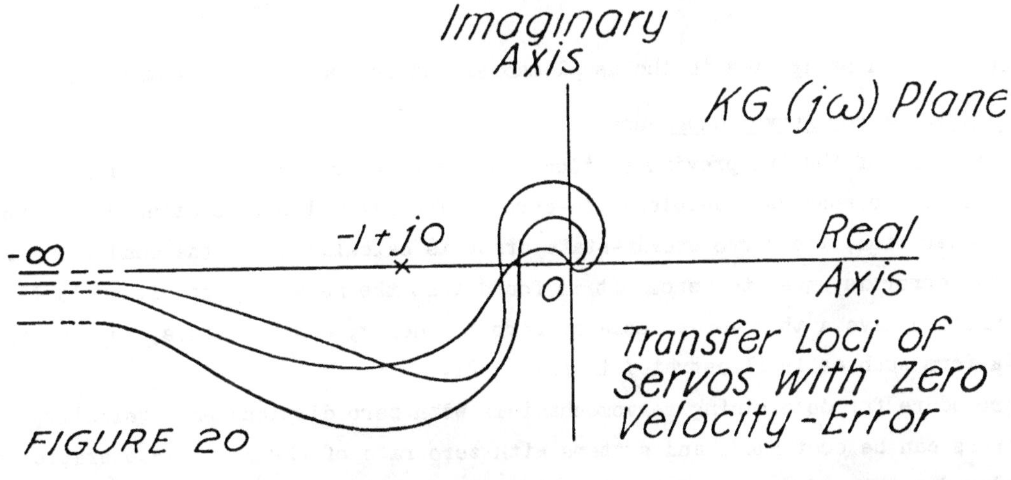

FIGURE 20

therefore, that must be fulfilled in order that a servo have zero velocity-error is that the transfer-function be of such form that its locus approaches infinity along the negative real axis as the frequency approaches zero.

Means of obtaining zero velocity-error servomechanisms are discussed in Chapter V. Obviously, a system with zero velocity-error also will have zero displacement-error.

Some explanation concerning the application of Nyquist's stability criterion to the transfer locus of figure 20 is necessary. At first it seems as if the locus with the unity point placed as shown to the left of the intersection with the real axis should be unstable, although in reality the system is stable. The explanation is that the solid-line curve, which is the locus for positive frequencies, and its conjugate, the locus for negative frequencies, do not join directly at zero frequency but are separated by 360°, since the first approaches -180°, while its conjugate approaches $+180^{\circ}$ as the frequency approaches zero. If the curves are considered as being connected by an infinite circle, as illustrated by figure 21, the Nyquist stability criterion can be applied in the usual manner and correct results obtained.

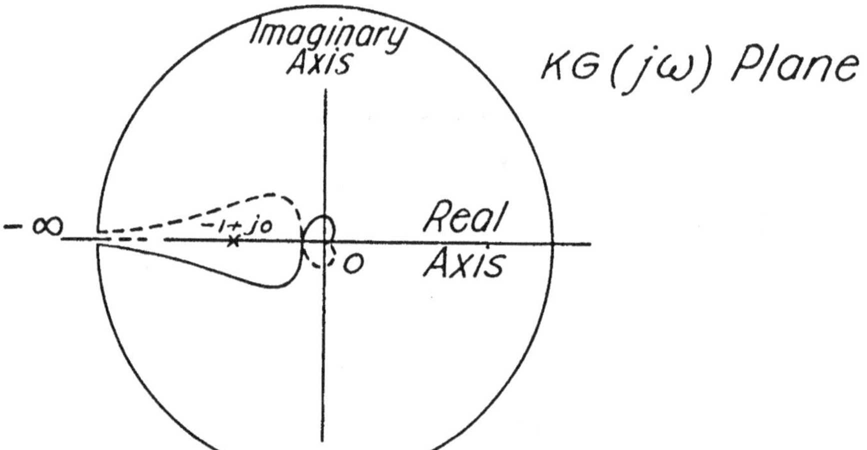

FIGURE 21

Zero Acceleration-Error Servomechanisms

The reasoning of the two previous sections can be extended and the criterion found for a servomechanism to possess zero acceleration-error. The servo input is assumed to consist of a constant acceleration, the servo steady-state error is calculated and the condition determined for which the error is equal to zero. It is found that the necessary condition is that the transfer-function have a third-order pole at zero frequency, and therefore, the transfer-locus will have a form such as is illustrated in figure 22.

The procedure for determining servomechanisms with zero displacement, velocity, and acceleration errors can be continued, and systems with zero rate of change of acceleration error, zero second order rate of change of acceleration error, etc., can be determined. It is possible to show that all these systems can be made stable and well-damped, although such systems are

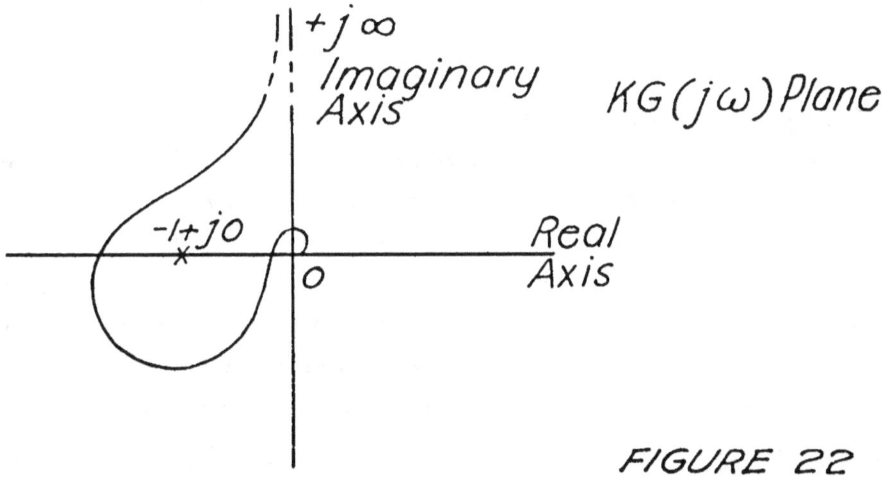

FIGURE 22

inherently slow, and transient response is sacrificed. Such systems should be designed only when warranted by the form of the servo input. Servomechanisms are very frequently subjected to inputs that are predominantly constant velocity, and it is generally necessary to so design the servo that the velocity-error is a minimum. Servomechanisms are less frequently subjected to such inputs that zero acceleration-error systems (or more complex systems) are required.

Shape of Transfer-Locus at Infinite Frequency

While the transfer-function of any physical system must approach zero as the frequency increases without limit, the manner with which the limit is approached is governed by the physical system involved. The general form of the transfer-function in terms of the complex variable, s, is given by Equation (82), and in terms of frequency, the complex variable, s, is replaced by jω. For large values of frequency, the transfer-function approaches

$$\underset{\omega \to \infty}{G(j\omega)} = \frac{(j\omega)^p}{(j\omega)^q} \qquad (91)$$

Both p and q are positive integers, and q must be larger than p if the servo system is physical Therefore, the transfer-locus approaches the zero point along the real or imaginary axis from a direction determined by the difference (q-p). For example, if (q-p) is equal to two, the transfer-locus approaches zero along the negative real axis; if (q-p) is equal to three, the transfer-locus approaches zero along the positive imaginary axis, and so on.

The difference (q-p), known as the "order" of the transfer-function, is a function of the number of cascaded coupled systems comprising the servo-controller and servo-motor. For example, suppose that the servo load consists of viscous damping and inertia; that the torque on the load is proportional to the displacement of a control motor characterized by inertia and damping, and that the torque on the control motor is proportional to the error. The equivalent circuit of this servo-controller and motor is two cascaded coupled circuits each of which adds two to the order of the transfer-function. Therefore, the total order (q-p) for the system is four and the transfer-locus approaches zero along the positive real axis. Physical circuits or devices cascaded in the servo system either increase the order of the transfer-function or leave it unaffected; no physical device cascaded in the system can decrease the order of the transfer-function.

The general shape of the transfer locus for most cases of interest can be sketched, since its form in three regions is known or can be determined easily. First, the shape of the transfer-locus at large frequencies is determined by the number of energy storage devices in the system. Second, the shape of the transfer-locus at small frequencies is determined by the imposed condition of no displacement-error, no velocity-error, etc. Third, the relation the transfer-locus must bear to the point (-1 + j0), if a stable system is to result, is known. This general knowledge of the transfer-locus is of particular value, since it enables many questions of servomechanism analysis to be resolved with little or no computation.

Ideal Zero Displacement-Error Servo

The simplest ideal displacement-error servo, if it could be realized physically, would be a system the transfer-function of which could be written,

$$K_b G_b(s) = \frac{K_b}{s} \tag{92}$$

The components of this system comprise an ideal servo motor the output velocity of which is instantaneously proportional to the displacement of the motor input, and a controller the output of which is proportional to the input (the servo-error) and independent of frequency.

The term, K_b, the gain of the system, is the product of the proportionality factor of the motor and that of the controller. This system cannot be realized physically, of course, if for no other reason than the order of the transfer-function, Equation (102), is unity while the order of the transfer-function of any physical system with a servo-motor load of finite mass, must be at least two. However, a brief consideration of this system is valuable becuase it reveals certain principles that are useful as guides in the design of physical systems.

The transfer-function of the ideal zero displacement-error servo as a function of frequency is

$$K_b G_b(j\omega) = \frac{K_b}{j\omega} \tag{93}$$

and the transfer-locus which lies along the negative imaginary axis of the complex plane, is illustrated by figure 23. The significant fact is that the relation of this transfer-locus to the point (-1 + j0) is independent of the gain, K_b , of the system; therefore, the shape of the amplitude and phase response of the servo output $\theta_o(j\omega)$, is independent of the system gain except for a frequency factor, as shown by figure 24. Evidently the ideal zero displacement-error system can be made as fast as desired, since the gain, K_b, can be set at any desired level, and the amplitude response of the output made to approximate unity and the phase response of the system made to approximate zero over as wide a frequency range as necessary.

This conclusion may be checked by calculating the servo error when a step input is applied to the system. If Equation (92) is substituted into Equation (29), the Laplacian transform of the error of the system is obtained:

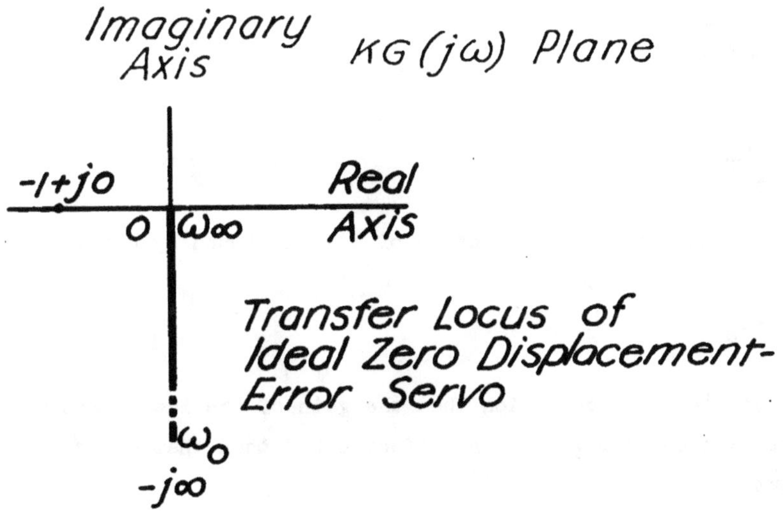
Imaginary Axis
KG(jω) Plane
-1+j0
Real Axis
0
ω∞
Transfer Locus of Ideal Zero Displacement-Error Servo
ω0
-j∞

FIGURE 23

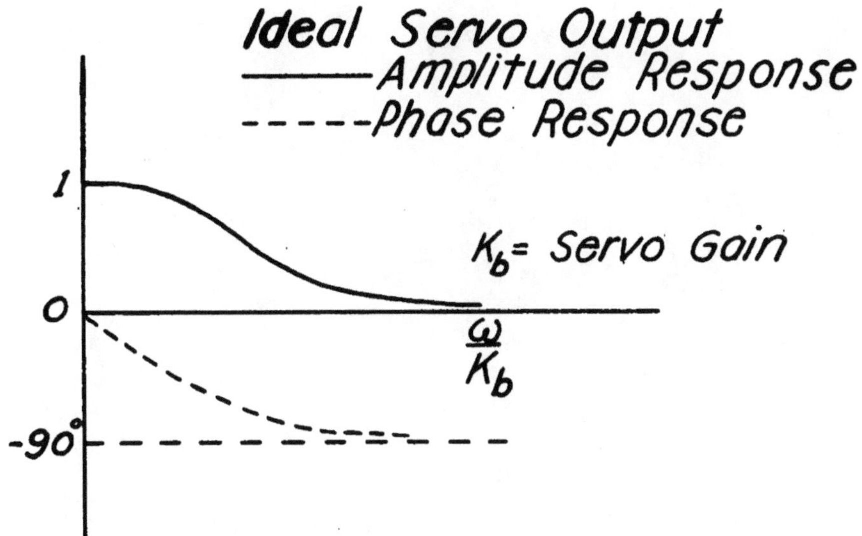
Ideal Servo Output
——Amplitude Response
-----Phase Response
1
0
-90°
Kb= Servo Gain
ω/Kb

FIGURE 24

$$E_{1b}(s) = \frac{1}{1 + \frac{K_b}{s}} \frac{1}{s} \tag{94}$$

$$E_{1b}(s) = \frac{1}{K_b + s} \tag{95}$$

The expression for the error as a function of time is found by obtaining the inverse transform of (39).

$$E_{1b}(t) = e^{-K_b t} \tag{96}$$

Equation (96) verifies the conclusion that the gain of the ideal zero displacement-error servo may be increased without limit, with no effect other than that of increasing the rate of decay of the transient.

CHAPTER IV

APPLICATION OF TRANSFER-LOCI TO THE ADJUSTMENT OF SERVOMECHANISMS

Servomechanism Design Prodecure

It is not possible to set up a formal, well-defined, system of servo design because of the individual nature of specific applications. However, a general procedure, each step of which is applied to almost every servo design problem, is as follows: 1) preliminary survey and selection of a servo system for investigation; 2) design of the basic components of the system in order to achieve optimum servo characteristics commensurate with economic considerations; 3) determination of the adjustable parameters of the system to obtain optimum servo performance; 4) design of additional devices to be added to the basic servo system for the purpose of compensating for failings in that system.

Certain servomechanism applications may demand a design procedure that follows the above outline rigorously; others may be such that certain of the steps receive only momentary consideration; almost every design problem, however, requires that each of the above steps be given some thought.

The selection of the basic system and the design and procurement of the components of that system (the first two steps in the design procedure) involve factors other than servo performance; such factors as cost, life, ease of manufacture, weight and size, are always important and may determine the basic servo system. These questions are answered only by engineering judgment and experience with servomechanism design.

The determination of the adjustable parameters of the basic system (step three) in order that optimum servo performance result is greatly aided by a study of the transfer-loci of the basic system. It is the purpose of this chapter to indicate how the transfer-loci may be so interpreted that useful results are obtained. After the servo system is in optimum adjustment, the decision must be made as to whether or not the system meets the demands of the application. This decision may be made on the basis of the amplitude response of the system as indicated in Chapter I, or upon the basis of the results of a transient analysis. If the performance of the system is inadequate to meet the requirements of the application, it may be worth while to investigate certain compensating devices for overcoming the failings of the system. The design of certain of these devices is dealt with in later chapters.

Factors Important in Servo Adjustment

The term "adjustment" as used in this paper denotes the general process by which those physical parameters of a given system that may be altered are determined. The term "adjustment" includes not only laboratory selection, but also design selection of the constants of the servomechanism.

The parameters of a basic system are generally adjusted with the following criteria as goals:

1. The servo must be stable.
2. The natural frequency or frequencies of the system should be high.

3. The damping of each natural frequency should be high.

4. The gain factor, "K", of the system should be high.

The above factors are all interdependant and except for the first, they are relative in as much as the term "high" actually means "as high as possible (and necessary) without undue sacrifice on the part of the other requirements." Therefore, no servo adjustments can be made without considering the effect upon all four factors.

It is difficult to calculate the mathematically optimum adjustment of all but very simple systems because of the complexity of the system and the difficulty of setting up performance criteria for specific applications. This was illustrated in Chapter II. The exact determination of the theoretical optimum servo adjustments is generally unnecessary, however, because of (1) the insufficient accuracy with which mechanical and electrical constants of most servo components can be specified, and (2) the necessity that the servo performance realized exceed the application requirements by an adequate margin of safety.

The accuracy with which the physical and electrical constants of servo systems can be specified frequently may not exceed twenty or thirty per cent. The necessary calculations and careful adjustments of the parameters of such systems in order to achieve mathematically optimum conditions are not warranted.

It is considered good practice for the actual servo performance to exceed the application requirements by an adequate margin of safety in order to allow for deterioration of the servo performance caused by wear and ageing, variations in the servo load, and other such factors. Because this margin of safety is generally considerable, servo adjustments, in most cases, should result in improvements in servo performance by a factor or the order of one and one-half in order to be justifiable. More strictly, the improvements obtained by adjustment should be an appreciable fraction of the margin of safety in order to be worth while. A workable procedure for many problems is to design the parameters of the system that they approach as closely as possible their optimum value, and make use of laboratory or factory adjustment for the final "tuning up."

It is not claimed that the methods explained in this and later chapters enable a system to be adjusted to its mathematically optimum response, but it has been found that in most cases the methods do predict the correct parameter values within about twenty per cent. This is sufficiently high accuracy for most applications.

Frequency Transformation

The study and interpretation of transfer-loci is aided by making a frequency transformation such as

$$\frac{\omega}{\omega_o} = u \tag{97}$$

in the transfer-function. The physical reason for this transformation is based upon the fact that in every system certain parameters are either entirely fixed or may be changed only with such difficulty that they can be considered fixed. For design purposes these parameters are considered invariant and the optimum value of the adjustable parameters determined in terms of the fixed parameters. This procedure, which is clarified by the consideration of an example in the latter part of this chapter, both simplifies the necessary calculation and enables the results to be applied more universally.

The frequency transformation, Equation (97), corresponds to a time transformation of the form

$$t' = \omega_o t \ . \tag{98}$$

The proof for this lies in the following theorem of Laplace transform theory:*

$$\mathcal{L}^{-1}\left[F\left(\frac{s}{\omega_o}\right)\right] = \omega_o\, f(\omega_o t), \tag{99}$$

if

$$\mathcal{L}^{-1}\left[F(s)\right] = f(t),$$

in which $\mathcal{L}^{-1}$ signifies "the inverse transform of."

Selection of the Gain Factor "K"

The particular servo parameter most easily adjusted is, perhaps, the gain factor "K", which is the frequency invariant portion of the transfer-function, $KG(j\omega)$. If the servo-controller is an electronic amplifier, the gain may be adjusted, in general, by regulating a voltage divider controlling the gain of the amplifier. The gain factor of systems incorporating controllers other than electronic amplifiers may be less easily adjusted, but the procedure in most cases is still simple. Since the gain is easily adjusted, it is important to develop a technique of gain adjustment that yields optimum results in servo performance.

The fourth of the factors listed on page 46 as guides to the proper adjustment of servomechanisms is that the gain "K" of the system should be high. The effects of increasing the gain are: (1) the velocity error of the system is reduced; (2) errors caused by restraining torques on the servo output are reduced; (3) errors caused by mechanical misalignment in the servo motor or controller are reduced; (4) the imaginary components of the complex roots of the system (the natural frequency or frequencies of the system) are increased in most cases; (5) the magnitudes of the real roots of the system are increased in most cases; and (6) the real parts of the complex roots of the system (the damping constants) in general are decreased. The first three of the above effects are reductions in the system steady-state error; the next two are increases in the speed of response of the system, and the last is a reduction in the speed of response of the system in that the time required for a transient oscillation to damp out is increased. As a matter of fact, continued increase of the gain factor of all physical servo systems results in servo instability. The general procedure, therefore, is to increase the gain of the system to the point beyond which further increments in gain result in a loss of servo performance.

The exact point at which the gain should be set depends upon the specific application for which the servo is designed. Certain applications require the damping ratios of the system to be large while others permit the damping ratios to be relatively low. A knowledge of the

* See reference 9, p.226

transient response of the system for various values of the gain factor enables the proper gain factor to be selected. Figure 5 illustrates the effect of various system gains upon the servo response to a step input. If the gain of the system is so low that all the roots of the characteristic equation (Equation (34)) of the system are real and unequal, the response of the system is illustrated by a curve such as curve A. As the system gain is increased, this particular system becomes in order, less overdamped, critically damped (curve B), underdamped with a large damping factor (curve C), underdamped with a small damping factor (curve D), and finally unstable. A response similar to that illustrated by curve C is close to the optimum for most servo applications.

A study of the servo response to a step function, therefore, is an effective means for determining the proper adjustment of the gain of the system. This information may be obtained from a transient analysis of the system or from direct observation. A transient analysis, except for simple systems, is laborious, but direct observation is a very effective means of adjusting servos for many applications. Laboratory arrangements for applying step inputs, or approximate step inputs, are usually very simple, and it is often only a matter of a few moments to observe the servo response for various values of gain and to choose that particular gain that results in optimum servo response as indicated by rapid transient decay.

A study of the amplitude response of the system also facilitates the proper selection of the gain. Amplitude responses of a typical system for a range of gain settings are illustrated by figure 25. As the gain is progressively increased, the response varies from that of curve A to that of curve D. The amplitude responses illustrated by figure 25 correspond to the similarly lettered transient response curves of figure 5. Thus curve C of figure 25 is the response of a system the gain of which is set close to the optimum value. In general, the gain can be increased until the peak in the amplitude response of $\frac{\theta_o}{\theta_i}(j\omega)$ is approximately one and one-third. The frequency at which this peak occurs is known as the resonant frequency of the system.

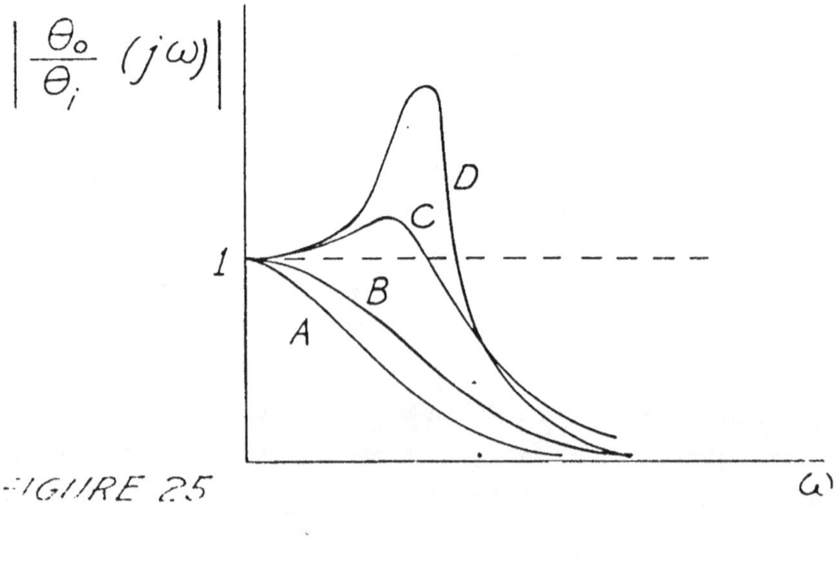

FIGURE 25

If the servo gain is to be adjusted in the laboratory, it is much simpler to do so on the basis of transient response rather than on the basis of sinusoidal response. If the correct gain is to be calculated, however, it is simpler to do so on the basis that the peak in the amplitude response should be approximately one and one-third, than that the transient response should be of a prescribed form. The reason for this is that the system gain that limits the maximum value of the amplitude response of $\frac{\theta_o(j\omega)}{\theta_i}$ to a prescribed value is easily found by graphical means from the plot of the transfer-locus of the system. Several methods for determining the gain are now considered.

A series of transfer-loci for a system the gain of which is varied is drawn in figure 26, and illustrates the fact that as the gain factor, K, of the transfer-function, $KG(j\omega)$, is increased, the loci approach more closely to the point $(-1 + j0)$. As was explained in ChapterIII, the value of $\left|\frac{\theta_o(j\omega)}{\theta_i}\right|$ for a particular frequency, ω_1, is the ratio of $|\overline{oc}|$ to $|\overline{ac}|$, in which the point c is the tip of the vector $KG(j\omega_1)$. This ratio is a maximum in the region in which the transfer-locus approaches most closely to the point $(-1 + j0)$. It is possible to plot a series of curves for various values of K, such as was done in figure 26, determine the maximum value

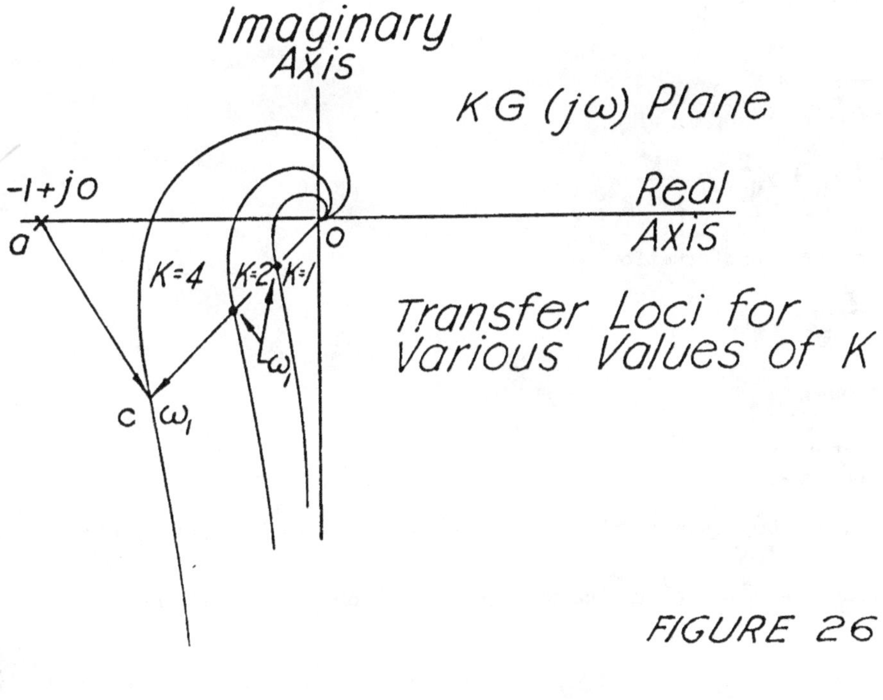

FIGURE 26

of $\left|\frac{\theta_o(j\omega_1)}{\theta_i}\right|$ for each locus by a cut-and-try process, and finally by interpolation find the value of K which limits the maximum of $\left|\frac{\theta_o(j\omega)}{\theta_i}\right|$ to a prescribed value. Such a procedure is unnecessarily lengthy and may be greatly simplified by recognizing that altering the gain factor of the system serves only to change the scale of the transfer-locus plot. This is illustrated by the loci of figure 26.

In order to make use of this fact in the selection of the proper system gain factor it is necessary first to show that the transfer-locus of a system in which $\left|\frac{\theta_o}{\theta_i}(j\omega)\right|$ is a constant has the form of a circle. This is shown as follows: Set the function $\left|\frac{\theta_{oM}(j\omega)}{\theta_i}\right|$ equal to a constant M.

$$\left|\frac{\theta_{oM}(j\omega)}{\theta_i}\right| = \left|\frac{K_M G_M(j\omega)}{1 + K_M G_M(j\omega)}\right| = M \tag{100}$$

Let the vector $K_M G_M(j\omega) = x_M + jy_M$ (101)

Substituting (101) into (100)

$$\left|\frac{\theta_{oM}(j\omega)}{\theta_i}\right| = \frac{\sqrt{x_M^2 + y_M^2}}{\sqrt{(1 + x_M)^2 + y_M^2}} = M, \tag{102}$$

$$\frac{x_M^2 + y_M^2}{(x_M + 1)^2 + y_M^2} = M^2, \tag{103}$$

$$x_M^2 + \frac{2M^2}{1 - M^2} x_M + y_M^2 = \frac{M^2}{1 - M^2} \tag{104}$$

Change variable by the transformation

$$x'_M = x_M + \frac{M^2}{1 - M^2} \tag{105}$$

Equation (104) becomes

$$x'^2_M + y_M^2 = \frac{M^2}{(M^2 - 1)^2} \tag{106}$$

The locus of Equation (106) is a circle the center of which is located at $x'_M = 0$, $y_M = 0$, and the radius of which is $\frac{M}{M^2 - 1}$. In the untransformed complex plane, the radius is the same but the center of the circle is located at $x_M = -\frac{M^2}{M^2 - 1}$, $y_M = 0$. The location and size of the circle are illustrated in figure 27.

Other useful relations involving the form and position of the circle of constant $\left|\frac{\theta_o(j\omega)}{\theta_i}\right|$ are:

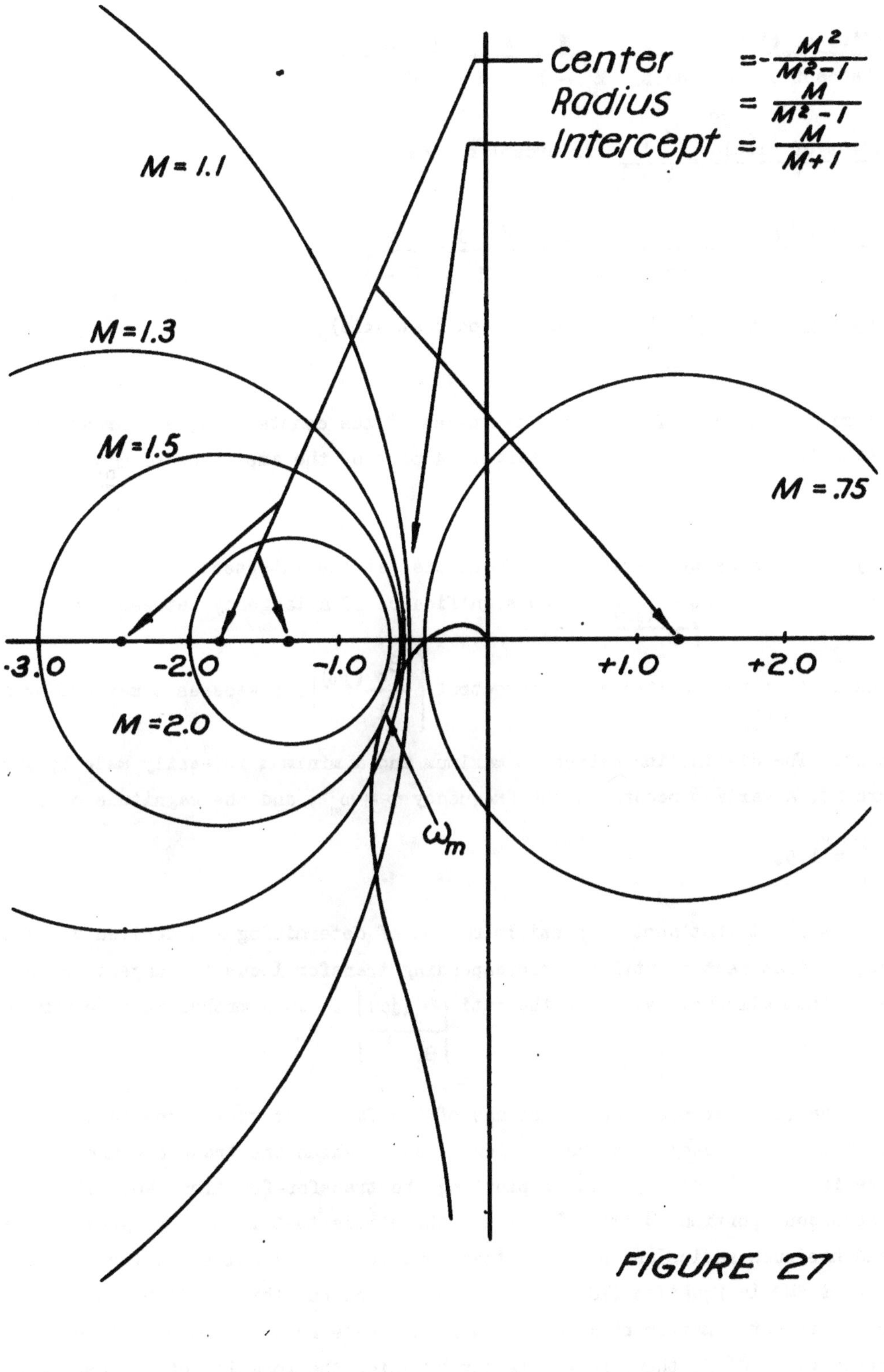
Center $= -\frac{M^2}{M^2-1}$
Radius $= \frac{M}{M^2-1}$
Intercept $= \frac{M}{M+1}$
M = 1.1
M = 1.3
M = 1.5
M = .75
-3.0
-2.0
-1.0
+1.0
+2.0
M = 2.0
ω_m

FIGURE 27

$$\frac{\textbf{radius of circle}}{\textbf{intercept on real axis}} = \frac{M}{M^2 - 1} \cdot \frac{M + 1}{M} = \frac{1}{M - 1} \tag{107}$$

$$\frac{\text{radius of circle}}{} = \frac{1}{M - 1} \text{ (intercept on real axis)}, \tag{107a}$$

$$\frac{\text{center of circle}}{\text{intercept on real axis}} = \frac{M^2}{M^2 - 1} \cdot \frac{M + 1}{M} = \frac{M}{M - 1} \tag{108}$$

$$\frac{\text{center of circle}}{} = \frac{M}{M - 1} \text{ (intercept on real axis)} \tag{108a}$$

A family of circles for particular values of the constant, M, are drawn in figure 27, together with the transfer locus of a servo. A plot of the amplitude of $\frac{\theta_{o(j\omega)}}{\theta_i}$ as a function of frequency can be made by reading the frequencies at the intersections of the transfer locus and the circles of constant $\left|\frac{\theta_{o(j\omega)}}{\theta_i}\right|$. The significance of a tangency between a circle and the transfer locus is that the amplitude of the output $\left|\frac{\theta_o}{\theta_i}(j\omega)\right|$, possesses a maximum or a minimum at that point. The distinction between a maximum and a minimum is easily made by inspection. Thus in figure 27, a maximum occurs at the frequency $\omega = \omega_m$, and the magnitude of the maximum is $\left|\frac{\theta_o(j\omega_m)}{\theta_i}\right| = 1.5$.

It is evident that another possible method of determining the desired value of gain factor is to adjust that factor until the corresponding transfer locus is tangent to the circle drawn for the maximum allowable value of the ratio $\left|\frac{\theta_o(j\omega)}{\theta_i}\right|$. This method is also unnecessarily lengthy. The simplest procedure makes use of the fact that variations in the gain, K, are equivalent to scale changes in the complex plane in which the transfer-function is plotted. The procedure is as follows: Instead of plotting the transfer-function, $KG(j\omega)$, plot only the frequency dependent portion, $G(j\omega)$. Construct the circle that is both tangent to the locus of $G(j\omega)$ and has such radius and position that Equation (108a) holds for the desired value of M. The value of "M" in Equation (108a) should be one and one-third. With the help of a pair of dividers it is but a matter of a few seconds to locate such a circle by a cut-and-try process. The correct value of K, then, is the factor by which the location of the center of this circle on the negative real axis of the $G(j\omega)$ plane must be multiplied in order that it equal $\frac{M^2}{M^2 - 1}$, which is the center of the circle in the $KG(j\omega)$ plane.

As an example, assume that the locus plotted in figure 28 is the locus of the frequency de-

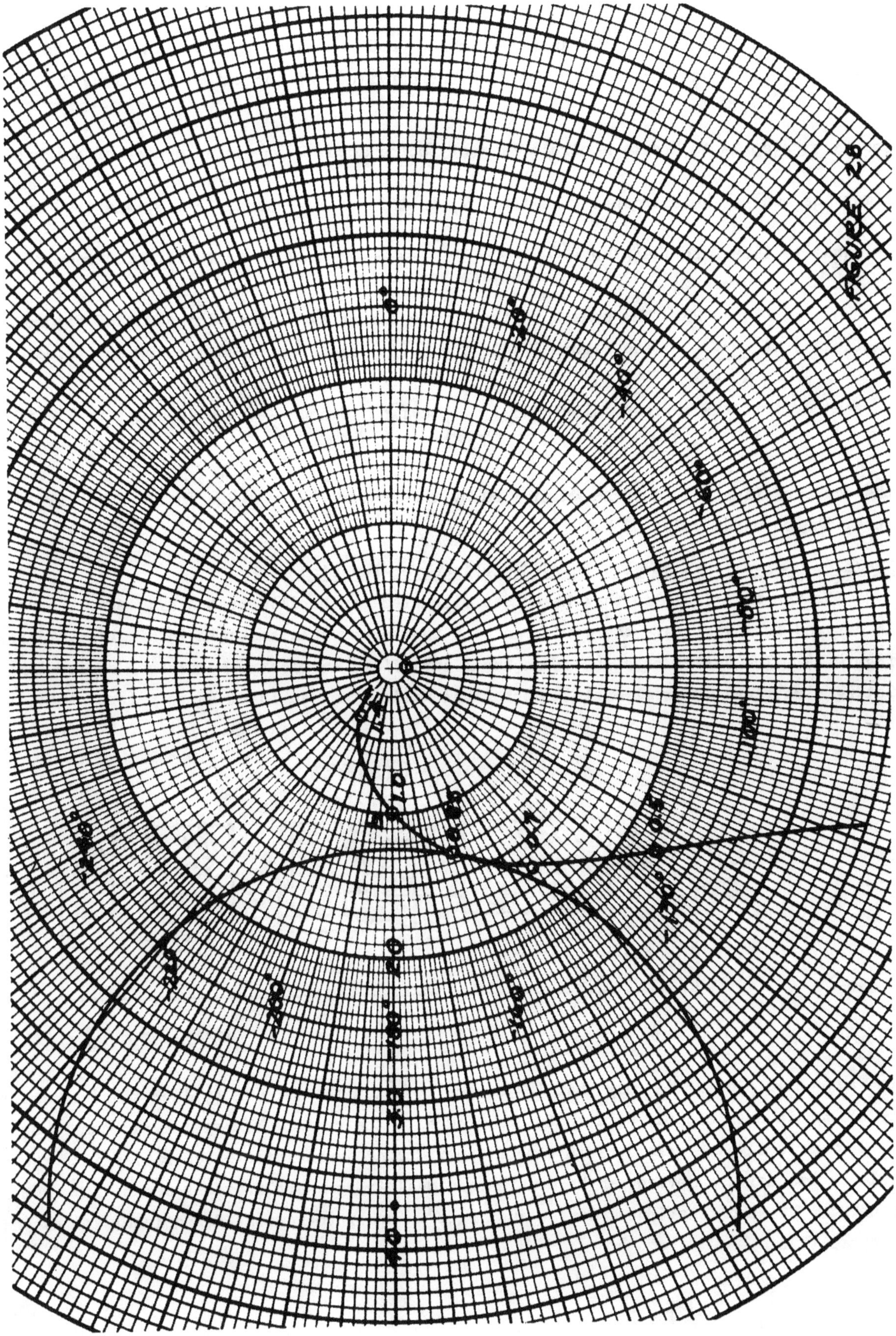

FIGURE 28

pendent portion $G(j\omega)$, of the transfer-function, and the value of the gain, K, is to be determined that will limit the maximum value of the amplitude of the output function, $\left|\frac{\theta_o(j\omega)}{\theta_i}\right|$, to 1.5.

The first step is to locate a circle tangent to the locus $G(j\omega)$, the center of which is $\frac{M}{M-1}$ times the real-axis-intercept. The term, M, in this case is equal to 1.5, and the factor $\frac{M}{M-1}$ is equal to three, so that the $\frac{\text{center of circle}}{\text{intercept on real axis}}$ is equal to three. The required circle is shown in figure 28. The center of the circle occurs at 3.68 on the scale of figure 28, but must occur at $\frac{M^2}{M^2-1}$ (see figure 27) in the $KG(j\omega)$ plane provided the correct value of K is chosen to limit the maximum value of $\left|\frac{\theta_o(j\omega)}{\theta_i}\right|$ to 1.5. Since $\frac{M^2}{M^2-1}$ is equal to 1.8 if M = 1.5, the scale of figure 28 must be multiplied by the factor $\frac{1.8}{3.68} = .49$. The factor 0.49 is, therefore, the required value of K.

Example: Servomechanism with Third-Order Transfer-Function

The design principles that have been discussed can be illustrated by a simple example. It is desired to investigate the performance of the servomechanism shown in diagrammatic form in figure 29. The system consists of an ideal amplifier, an electro-mechanical device for translating electrical information into mechanical motion, and an ideal servo motor. What is meant by

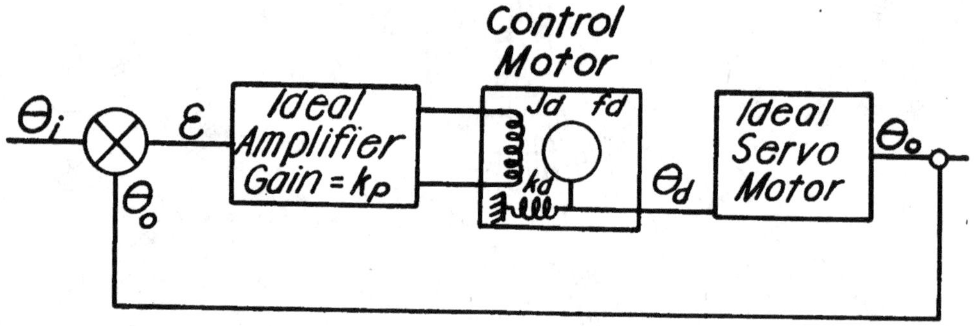

FIGURE 29

the adjective "ideal" is only that the device to which the term is applied follows a particular relationship sufficiently well that, compared with the rest of the system, it can be assumed to follow perfectly the relationship without introducing appreciable inaccuracies into the final result. Obviously, a device can be considered ideal in some applications and be far from ideal in others. The specifications a device must meet in order to be considered ideal are discussed in Chapter VII.

The servo motor of figure 29 is a motor whose output velocity is instantly proportional to the input displacement. An hydraulic system consisting of a variable displacement pump linked hydraulically to a fixed displacement motor might approximate such a servo motor. The input displacement of the motor of figure 29 corresponds to the displacement from neutral of the variable displacement pump, and if leakage in the hydraulic system is negligible, the velocity of the hydraulic motor, corresponding to the output velocity of the servo motor, is proportional to that displacement. Provided the inertia coupled to the hydraulic motor shaft and the leakage in the hydraulic system are sufficiently small, the time required for the servo output to reach its final velocity when the input is given a sudden displacement can be neglected in comparison with the time lags present in the remainder of the servo system, and the servo motor can be considered ideal. Stated another way, the servo motor can be considered ideal provided that, when considered alone, its response is well-damped and has a natural frequency that is several times that of the control motor.

The control motor driving the servo motor is characterized by a moment of inertia, J_d, viscous damping, f_d, and an elastic restraining force, k_d. The torque of the control motor is assumed to be directly proportional to the amplifier output which in turn is related by a constant to the amplifier input. The equations of the system are the following:

$$T(s) = k_p E(s) \tag{109}$$

(The torque of the control motor is proportional to the amplifier input.)

$$T(s) = (J_d s^2 + f_d s + k_d)\,\theta_d(s) \tag{110}$$

(The equation of motion for the control motor)

$$s\,\theta_o(s) = k_m \theta_d(s) \tag{111}$$

(The output velocity of the servo motor is proportional to the input displacement of the servo motor.)

In the above equations,

$T(s)$ = transform of the torque developed by the control motor.

$\theta_d(s)$ = transform of the control motor displacement and input to the servo motor.

k_p = amplifier and control motor constant.

k_m = servo motor constant.

The transfer-function of the third-order system, $K_3G_3(s)$ is

$$K_3G_3(s) = \frac{\theta_o(s)}{E(s)} = \frac{k_p k_m}{s(J_d s^2 + f_d s + k_d)} \tag{112}$$

If the following substitutions

$$\frac{k_d}{J_d} = \omega_o^2 , \tag{113}$$

$$\frac{f_d}{J_d} = 2\zeta_3\omega_0 , \tag{114}$$

$$\zeta_3 = \frac{1}{2}\frac{f_d}{\sqrt{k_d J_d}}, \tag{114a}$$

are made in Equation (112), the transfer-function becomes

$$K_3G_3(s) = \frac{k_p k_m}{J_d} \frac{1}{s(s^2 + 2\zeta_3\omega_0 s + \omega_0^2)} \tag{115}$$

The quantity ζ_3 is known as a damping ratio. If the transformation

$$\frac{s}{\omega_0} = s' , \tag{116}$$

is applied to Equation (115), Equation (117) is obtained.

$$K_3G_3(s') = \frac{k_p k_m}{\omega_0 k_0} \frac{1}{s'(s'^2 + 2\zeta_3 s' + 1)} \tag{117}$$

The transfer-function is obtained in terms of a transformed frequency operator, ju, by substituting that operator for the complex variable, s' , in Equation (117).

$$K_3G_3(ju) = \frac{k_p k_m}{\omega_0 k_d} \frac{1}{ju(-u^2 + 2j\zeta_3 u + 1)} \tag{118}$$

The transformed frequency, u, is related to the true angular frequency, ω, by

$$\frac{\omega}{\omega_0} = u, \tag{119}$$

in accordance with the corresponding Equation (116).

The procedure that has resulted in the form of Equation (118) for the transfer-function is a frequency transformation of the original expression (56) using the undamped natural frequency of the control motor, $\omega_0 = \sqrt{\frac{k_d}{J_d}}$ as a measure of frequency. The approach follows the general method outlined on p. 47 and the consistency of transformations (113), (114), and (116) can be quickly proven by a dimensional analysis of these relations. The philosophy underlying the choice of the ratio $\sqrt{\frac{k_d}{J_d}}$ as the frequency base is the following: The performance of the servo system always may be improved by increasing the natural frequency, ω_0, of the control motor, and it is desirable, therefore, to have this frequency as high as practical. The natural frequency, ω_0, is determined by the ratio of the maximum torque of the control motor to the motor moment of inertia (known as the torque-to-inertia-ratio), and the maximum required deflection of the motor; thus, if θ_{dm} is the maximum required deflection of the motor and T_m is the maximum available torque of the motor,

$$\theta_{dm} = \frac{T_m}{k_d} \qquad (120)$$

$$\omega_0 = \sqrt{\frac{k_d}{J_d}} = \sqrt{\frac{T_m}{J_d}\,\frac{1}{\theta_{dm}}} \qquad (121)$$

in which $\frac{T_m}{J_d}$ is equal to the torque-to-inertia ratio of the motor. Both the maximum required deflection, θ_{dm} , and the torque-to-inertia ratio are relatively fixed quantities; the first is fixed by design and the second by the geometrical configuration and principle of operation of the control motor. While the torque-to-inertia ratio should be large and the maximum deflection, θ_{dm}, small for best servo operation, both quantities are primarily chosen from such standpoints as ease of manufacture, cost, and the like. The proper procedure, therefore, is to secure the optimum servo performance for given values of torque-to-inertia ratio and maximum deflection, θ_{dm} . If the optimum servo performance is inadequate, the alternatives are to increase the torque-to-inertia ratio of the control motor, reduce the maximum required deflection, θ_{dm}, or to incorporate compensating devices in the servo system. By Equation (121) the frequency base, ω_0, is a function of the torque-to-inertia ratio and the maximum deflection of the control motor. The frequency transformation, (119), allows the other parameters and the resulting performance of the servo to be calculated in terms of the value of this frequency base.

Equation (118) is written as a product of two terms: one term that is a function of the frequency, ω , and the other that remains invariant with frequency. The frequency-invariant term is equal to the gain factor, K_3 , and the frequency dependent term is equal to $G_3(ju)$. Thus

$$K_3 = \frac{k_p k_m}{\omega_0 k_d} \qquad (122)$$

$$G_3(ju) = \frac{1}{ju(-u^2 + 2j\zeta_3 u + 1)} \qquad (123)$$

The problem is so to select the gain factor, K_3, and the damping ratio, ζ_3, that optimum servo performance is obtained. The solution is obtained by plotting the loci $G_3(ju)$ for various values of ζ_3 and so adjusting the gain factor, K_3, that for each locus the maximum value of $\left|\frac{\theta_o(ju)}{\theta_i}\right|$ does not exceed a preset value. The various loci are compared with one another and the one selected with which the criteria listed on p. 45, 46 are best met.

The loci for the third-order transfer-function servo are plotted in figures 30 and 31. The loci correspond to damping ratios ranging from 0.3 to 2.0, and the proper system gain factor can be determined for each damping ratic by employing the method explained in the preceding section.

Before choosing the optimum locus, it is of interest to determine the stability criterion

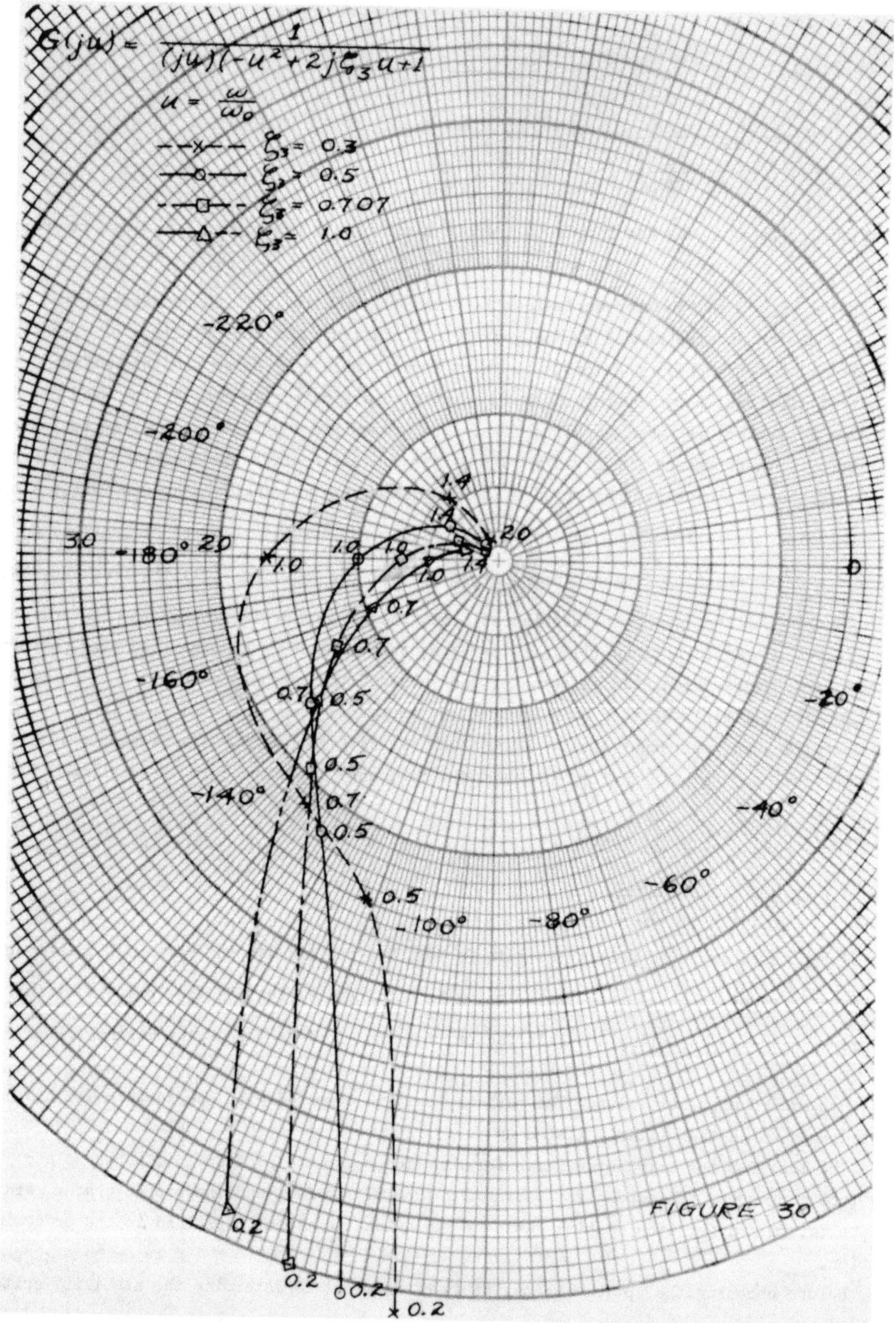

G(ju) = 1 / (ju)(-u² + 2jζ₃u + 1)
u = ω/ω₀
ζ₃ = 0.3
ζ₃ = 0.5
ζ₃ = 0.707
ζ₃ = 1.0
-220°
-200°
-180°
-160°
-140°
-100°
-80°
-60°
-40°
-20°
0
3.0
2.0

FIGURE 30

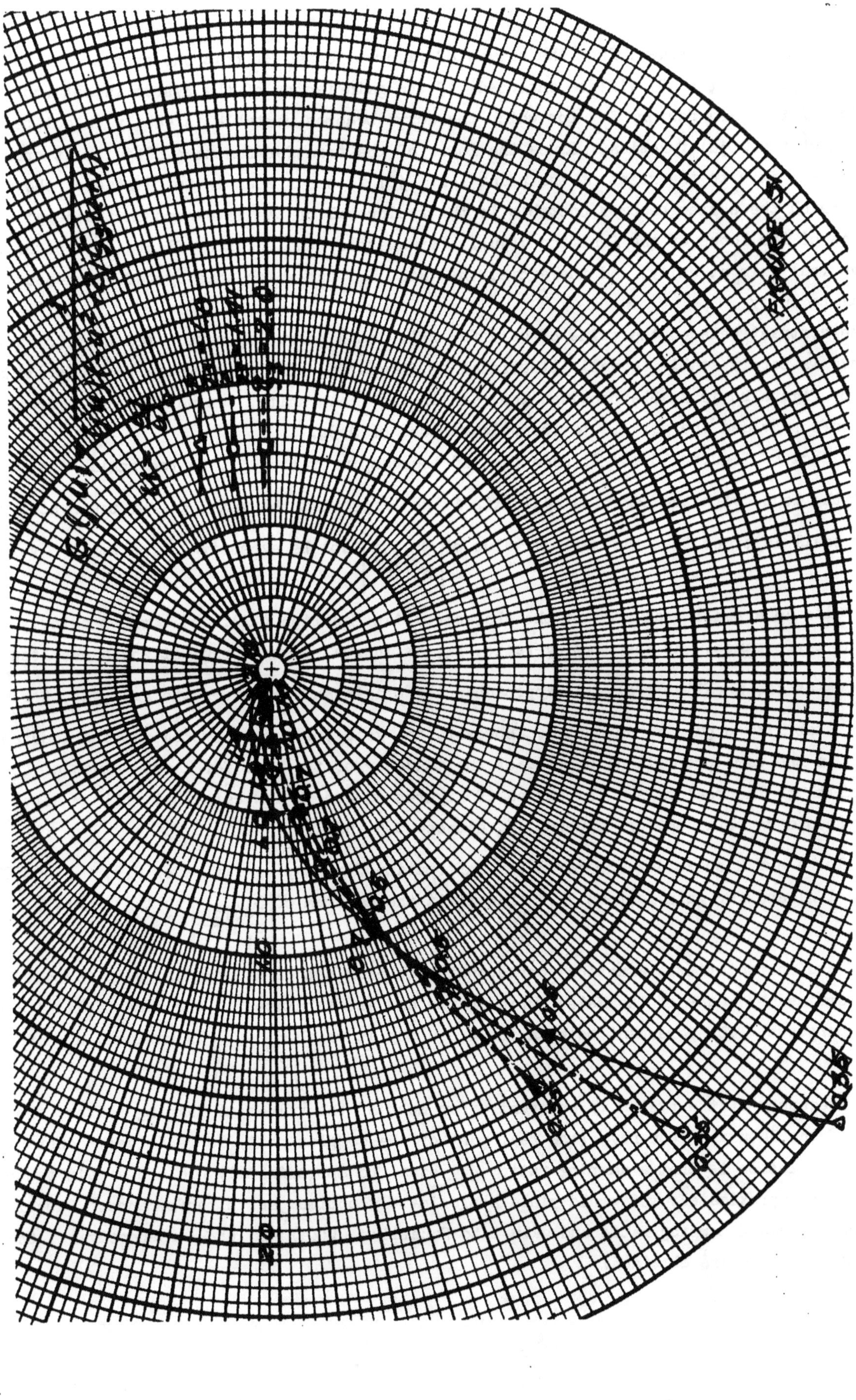

of the system. It is evident from the curves of figures 30 and 31 that the system will be unstable (the point $(-1 + j0)$ will be enclosed) if the gain factor of the system is set too large. The value of the gain factor, K_u, for which the servo becomes unstable is easily calculated from the locus of the $G(ju)$ function. Since the locus of the transfer-function $KG(ju)$ must cross the negative real axis to the right of the point $(-1 + j0)$ if the system is to be stable, the gain, K_u, for which the system becomes unstable is the reciprocal of the intercept on the negative real axis of the locus of the function $G(ju)$. For example, the locus of figure 28 crosses the negative real axis at unity, and therefore, the gain, K_u, for which this system becomes unstable is equal to unity.

The contrast between the value of gain for well-damped servo operation and that for which the servo is barely stable is well illustrated by figure 28. The servo whose locus is illustrated by figure 28 is stable provided the gain is less than unity, but well-damped operation only occurs provided the gain is in the neighborhood of 0.34.

The loci of figures 30 and 31 show that if the damping ratio, ζ_3, is too small, the intercept of the locus with the negative real axis is so large that the gain, K_3, of the system must be very small in order that the maximum value of $\left|\frac{\theta_o(j\omega)}{\theta_i}\right|$ not be excessive. Too small a value of gain results in a small real root of the characteristic Equation (34), and correspondingly poor transient performance. As the damping ratio, ζ_3, is increased, the point at which the locus of $G(ju)$ crosses the negative real axis approaches the origin and therefore the gain, K_u, for which system instability occurs is augmented. However, the gain for proper system damping does not increase in the same proportion, because of the tendency of the loci for large values of ζ_3 to lie parallel to the real axis. The curves of figures 30 and 31 indicate that optimum servo operation should be obtained for values of ζ_3 between 0.5 and 1.0.

Tables 1 and 2 summarize the pertinent information for damping ratios of 0.3, 0.5, 0.707, 1.0, and 1.41 and for gain factors "K_3" that limit the maximum of $\left|\frac{\theta_o(ju)}{\theta_i}\right|$ to 1.33 and 1.5. The roots of the characteristic equation corresponding to each set of damping and gain factors have been determined by means of the Weiss[24] cubic charts and are tabulated in the same tables for comparison purposes. The agreement between the frequency at which $\left|\frac{\theta_o(j\omega)}{\theta_i}\right|$ is maximum, u_m, and the natural frequency of the system is apparent. For most of the values of damping ratios, the damping of the natural frequency, (the real part of the complex root) is relatively constant.

The conclusion is that a damping ratio, ζ_3, equal to approximately 0.6 and a gain factor, K_3, equal to approximately .35 would yield approximately optimum servo performance. Substituting these values of damping ratio and gain into Equations (114a) and (122) respectively, the following relations are obtained:

$$\frac{f_d \omega_0}{k_d} = 1.2 \tag{124}$$

$$\frac{k_p k_m}{\omega_0 k_d} = .35 \tag{125}$$

Equations (124) and (125) permit the physical circuit constants to be calculated if the moment of inertia, J_d, and the maximum required deflection, θ_{dm}, of the control motor are known. Also, they show the manner in which the damping, f_d, and the amplifier gain, k_p, must be varied if optimum servo adjustment is to be maintained while either the moment of inertia, J_d, or elastic constant, k_d, of the control motor is varied.

The facts that the maximum value of $\left|\frac{\theta_{o(ju)}}{\theta_i}\right|$ corresponding to the above selected values of damping ratio and gain occurs at u equal approximately to 0.75, (which is, therefore, approximately the natural frequency of the system) and that the response of the system is given approximately by curve C of figure 5, allow the speed of response of the system to be calculated roughly, and the value of ω_0 to be chosen to meet specifications of a particular application.

The principles involved in the analysis and adjustment of any physical servo system are exactly the same as those applied in this example. While the above example is the analysis of a system having an ideal amplifier and servo motor and a motor for controlling the servo motor characterized by moment of inertia, viscous damping, and an elastic restraining force, the same conclusions hold for any system that has a transfer-function of the form given by Equation (115). Nichols[20] has shown that an hydraulic transmission servo motor can be characterized by the transfer function of Equation (115), so that if the rest of the system containing such a motor can be considered ideal, the same conclusions directly apply.

ζ_3	Determined from Transfer-Function Loci: Maximum Stable Gain K_u	Determined from Transfer-Function Loci: Gain Factor $\left\lvert \frac{\theta_o}{\theta_i}(ju) \right\rvert_{max} = 1.33$	Determined from Transfer-Function Loci: Resonant Frequency (u_m) $\left\lvert \frac{\theta_o}{\theta_i}(ju) \right\rvert_{max} = 1.33$	Determined from Cubic Charts $\left\lvert \frac{\theta_o}{\theta_i}(ju) \right\rvert_{max} = 1.33$: Real Root	Real Part of Complex Root	Imaginary Part of Complex Root
.3	.6	.32	.90	.36	.12	.94
.5	1.0	.46	.80	.44	.27	.83
.707	1.414	.51	.65	.77	.32	.63
1.0	2.0	.58	.60	1.56	.21	.51
1.41	2.83	.50	.40	2.4	.16	.32

Table I.

ζ_3	Determined from Transfer-Function Loci: Gain Factor $\left\lvert \frac{\theta_o}{\theta_i}(ju) \right\rvert_{max} = 1.5$	Determined from Transfer-Function Loci: Resonant Frequency (u_m) $\left\lvert \frac{\theta_o}{\theta_i}(ju) \right\rvert_{max} = 1.50$	Determined from Cubic Charts $\left\lvert \frac{\theta_o}{\theta_i}(ju) \right\rvert_{max} = 1.5$: Real Root	Real Part of Complex Root	Imaginary Part of Complex Root
.3	.33	.92	.32	.12	.95
.5	.49	.80	.62	.18	.85
.707	.58	.70	.98	.22	.74
1.0	.67	.65	1.63	.18	.60
1.41	.58	.45	2.5	.14	.45

Table II

CHAPTER V

THEORY OF MINIMUM VELOCITY-ERROR SERVOMECHANISMS: PART I

Zero Velocity-Error Systems

The restrictions that must be met by the transfer-function of a servo system if that system is to have zero velocity-error are developed in Chapter III. Means by which transfer-functions with zero velocity-error characteristics can be physically realized or approximated are discussed in this chapter.

It was shown in Chapter III that a servomechanism must have a transfer-function with a second-order pole at zero frequency if that system is to have no velocity error. It has also been shown that the transfer-function of a servo motor of the integrating type, that is, a motor the output velocity of which is proportional to its input displacement, has a pole of the first order at zero frequency. Hydraulic transmission systems have been cited as examples of integrating motors. If it is desired, therefore, to obtain a servo whose transfer-function has a second order pole at zero frequency, the first system that might be considered is one in which are cascaded two devices with single-order poles at zero frequency, such as illustrated in figure 32. Such a system will be shown to be inherently unstable, however, and should be avoided except under special circumstances.

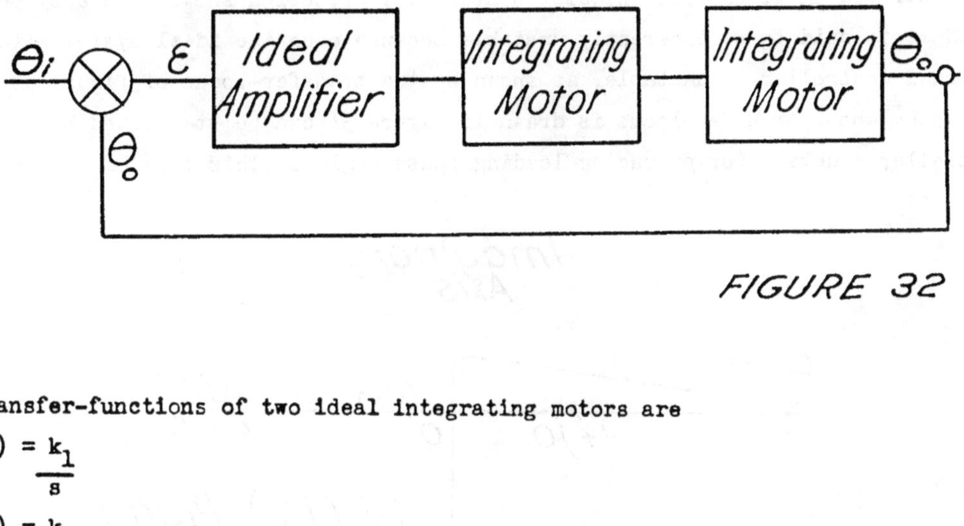

FIGURE 32

The transfer-functions of two ideal integrating motors are

$$K_1G_1(s) = \frac{k_1}{s} \tag{126}$$

$$K_2G_2(s) = \frac{k_2}{s} \tag{127}$$

The combined transfer-function of a system composed of two such motors in cascade is obtained by forming the product of (126) and (127).

$$K_TG_T(s) = K_1G_1(s)\ K_2G_2(s) = \frac{k_1k_2}{s^2} \tag{128}$$

$$K_TG_T(j\omega) = \frac{-k_1k_2}{\omega^2} \tag{129}$$

The locus of Equation (129) is a straight line extending along the negative real axis from zero to infinity as shown in figure 33, and passing directly through the point (-1 + j0).

The transfer-function of figure 33 is that of an idealized system and does not correspond to any physical system. The transfer-function locus of a physical servo motor of the integrating type is shifted by some varying negative angle from the locus of an ideal motor by the lag

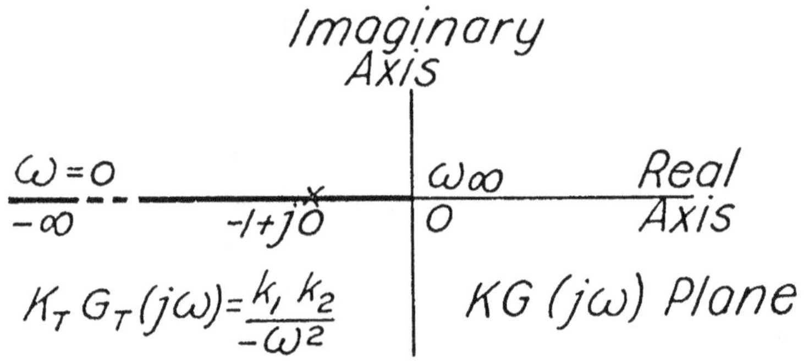

FIGURE 33

introduced by the mass of the motor output shaft. Any additional device in the system, such as the control motor discussed in the preceding section, introduces further lags in the system, and the negative shift of the transfer locus is increased. Therefore, the transfer-function locus of a physical system composed of two cascaded servo motors of the integrating type is of a form such as illustrated in figure 34. Nyquist's stability criterion shows such a system to be unstable, and it is said to be inherently unstable because even the ideal system with no lags and a proportional controller is unstable, as shown by the transfer-locus of figure 33.

The system whose transfer-locus is drawn in figure 34 can be stabilized by including in the servo-controller a device for producing leading phase angles. This type of device, discussed in

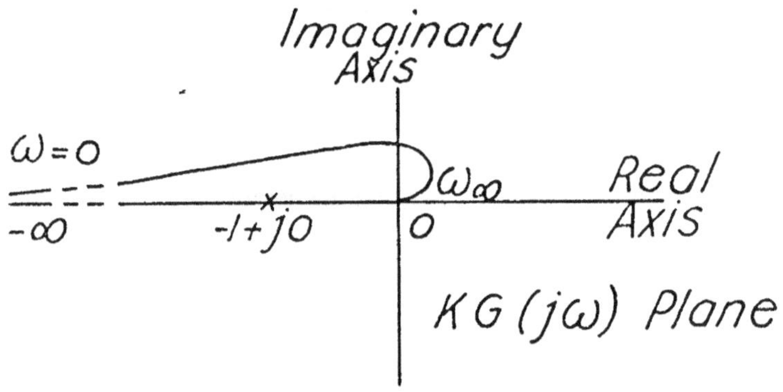

FIGURE 34

Chapter VI, leads to a system transfer-function locus of the type illustrated in figure 35. If the gain of such a system is so adjusted that the negative unity point lies within the portion of the loop shifted across the negative real axis the system is stable, but if the gain is too large or too small, the system is unstable. Unless warranted by special circumstances, an in herently unstable servo system should be avoided, and other means should be employed to secure a zero velocity-error servo.

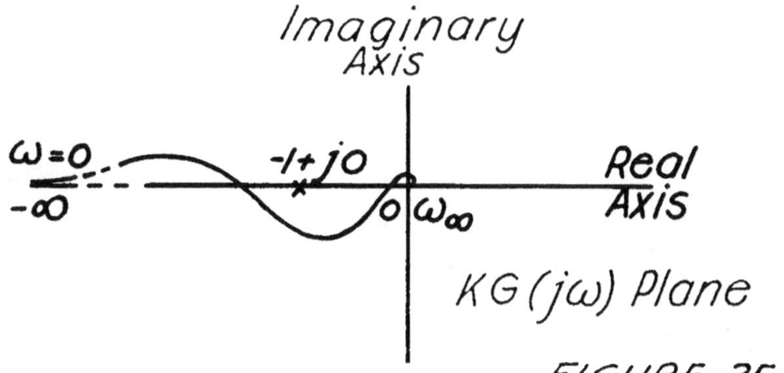

FIGURE 35

A more sound approach to the problem of obtaining zero velocity-error systems is to employ a servo motor of the integrating type and a servo-controller so designed that its response to very low frequencies approximates that of an integrating motor, but at high frequencies is small

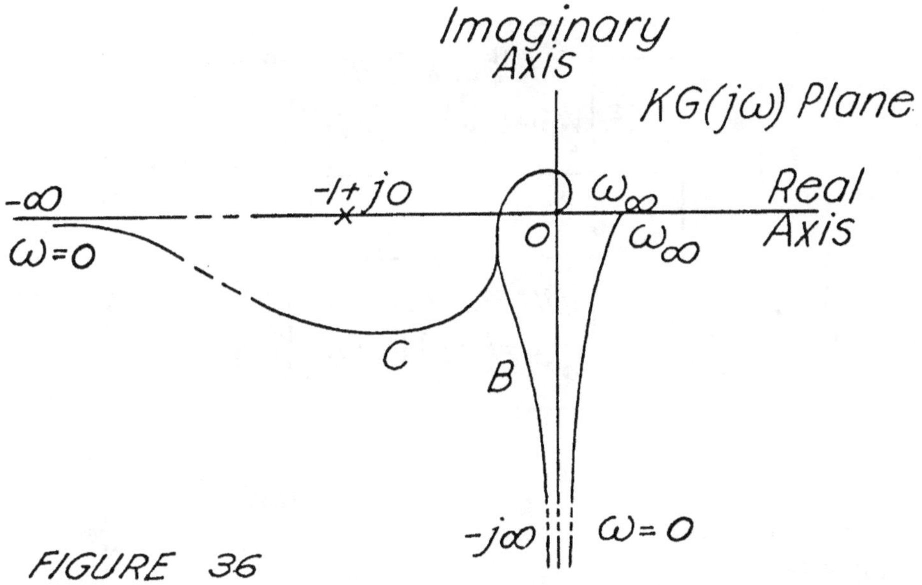

FIGURE 36

and constant, as illustrated in figure 36, curve A. In particular, the response at zero frequency of the controller must be infinite in order that the complete servo system have a second-order pole at zero frequency. If the controller transfer-function at all but very low frequencies is constant, the controller introduces phase-shift into the system at low frequencies only, and the resultant transfer-function of the complete system will have the form illustrated in figure 36, curve C, provided the transfer-function locus of the servo system without the compensating controller is illustrated by curve B, figure 36. If the gain of the complete servo is so adjusted that the intersection of its transfer-locus with the negative real axis lies to the right of the point (-1 + j0), as is the case with curve C, figure 36, the system is stable. The response of the system will be satisfactory provided the locus does not approach the point (-1 + j0) so closely that the maximum values of $\left|\frac{\theta_o(j\omega)}{\theta_i}\right|$ are excessive.

The preceding analysis has shown that well-damped servo operation combined with zero velocity-error may be expected provided that a controller having a transfer-locus of the general form illustrated by curve A, figure 36, can be physically realized. A physical device is required with the properties of infinite gain at zero frequency and very high gain at very low frequencies, but with a gain that is small and essentially constant at moderate and high frequencies. The fact that infinite gain is necessary indicates that the only type of device that can meet these requirements is an active, regenerative system, whose feedback path is so designed that the device has infinite gain at zero frequency but a finite, essentially constant gain above some predetermined frequency.

The block diagram of a regenerative controller with which the preceding requirements can be met is illustrated in figure 37.

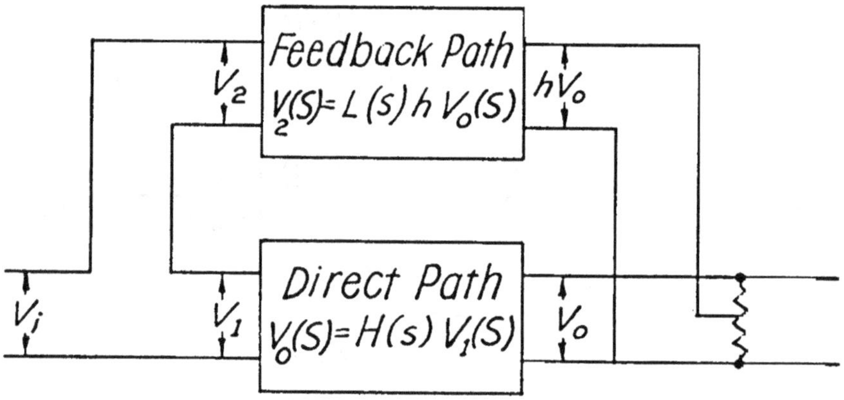

FIGURE 37

The equations of the controller are

$$V_o(s) = H(s)\ V_1(s) \tag{130}$$

$$V_2(s) = hL(s)\ V_o(s) \tag{131}$$

$$V_1(s) = V_i(s) + V_2(s) \tag{132}$$

in which

$V_1(s)$ = transform of input to controller,

$V_o(s)$ = transform of output of controller,

h = fraction of controller output fed into feedback path.

H(s) = transfer-function of direct path in controller,

L(s) = transfer-function of feedback path.

From Equations (130), (131), and (132) the transfer-function of the controller, $K_1G_1(s)$, is is found to be

$$K_1G_1(s) = \frac{H(s)}{1 - hH(s)\,L(s)} \tag{133}$$

One or both of the functions H(s) and L(s) must be frequency dependent in order to realize the desired characteristics. Assume that H(s) is independent of frequency and that the feedback path contains the frequency-dependent networks. In particular, let

$$H(s) = 1 \text{ (constant)}. \tag{134}$$

Then

$$K_1G_1(s) = \frac{1}{1 - hL(s)} \tag{135}$$

In terms of frequency the transfer-function of the controller is

$$K_1G_1(j\omega) = \frac{1}{1 - hL(j\omega)} \tag{136}$$

The transfer-function $K_1G_1(j\omega)$ is of the desired form provided that the function $L(j\omega)$ varies

approximately as indicated in figure 38. Either the magnitude of h or the maximum value of $L(j\omega)$ must be so adjusted that the gain of the controller is infinite at zero frequency, and the network in the feedback path is adjusted to control the rate at which $L(j\omega)$ decreases with increasing frequency, which, in turn, determines the manner in which the response of the controller

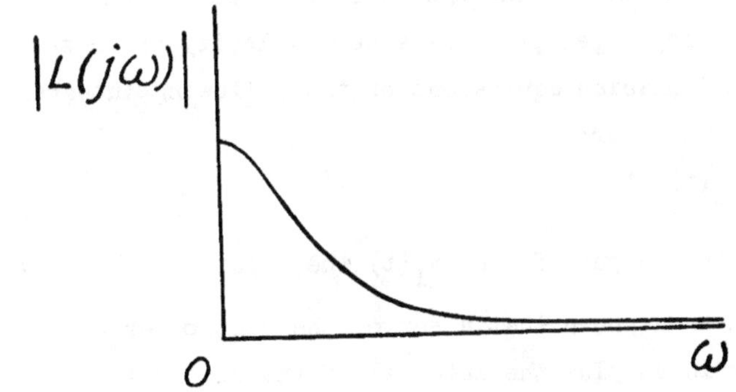

FIGURE 38

shifts from a large value at low frequencies to a small, constant, value at high frequencies.

The schematic of a simple type of network having a transfer-function $L(j\omega)$ of a form similar to that of figure 38 is drawn in figure 39. The transfer-function of this network is

$$\frac{V_2(s)}{hV_o(s)} = L(s) = \frac{1}{1 + RCs} \tag{137}$$

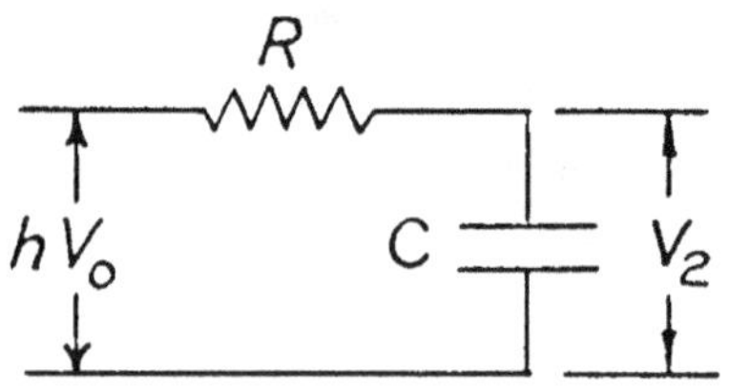

FIGURE 39

Substituting (137) into (135)

$$K_1G_1(s) = \frac{1}{1 - \frac{h}{1 + RCs}} \tag{138}$$

$$K_1G_1(s) = \frac{(1 + RC\ s)}{1 + RC\ s - h} \tag{138a}$$

$K_1G_1(s)$ has a pole of first order at $s = 0$ provided that $h = 1$.

Thus,

$$K_1G_1(s) = \frac{(1 + RCs)}{RCs} \tag{139}$$

Equation (139) may be written in the form

$$K_1G_1(s) = 1 + \frac{1}{RCs} \tag{139a}$$

The preceding reasoning has led to the synthesis of a servo-controller which, if used with a servo motor of the integrating type, produces a zero velocity-error servo system.

Equation (139a) is the Laplacian equivalent of the following integral equation specifying the response of the servo-controller:

$$v_o(t) = v_1(t) + \frac{1}{RC} \int v_1(t)\, dt \tag{140}$$

In Equation (140) $v_o(t)$ is the output of, and $v_1(t)$ the input to, the servo-controller, as functions of time. Thus it has been shown that a servo, the controller of which has a response proportional to the controller input plus the integral of the controller input, possesses zero velocity-error. This fact has been known for some time; it was first shown by Minorsky.[18] Its proof in this paper is incidental in the development of methods for synthesizing physical devices that will enable zero velocity-error servomechanisms to be realized. A servo-controller with the response specified by Equation (140) is frequently termed an integral-controller.

In terms of frequency, Equation (139a) is

$$K_1G_1(j\omega) = 1 + \frac{1}{jRC\omega} \tag{141}$$

The locus of (141) in the complex plane is illustrated by figure 40. At zero frequency the response of the controller has infinite magnitude and ninety degrees phase-shift. At infinite frequency the response has unit magnitude and zero phase-shift. The rate at which the transition

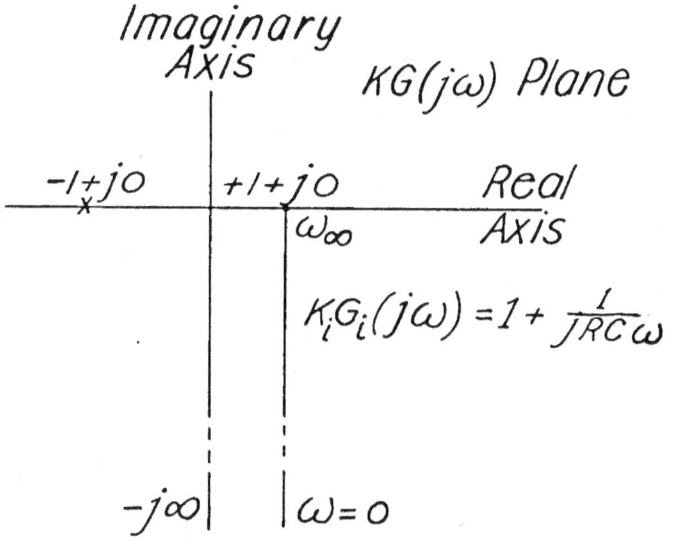

FIGURE 40

from infinite to unit magnitude occurs is controlled by the RC product of the controller feedback circuit.

The effect of varying the RC product of a servo with an integral-controller is determined by plotting the transfer-loci of the servo for various values of the RC product. For the purpose of discussion, assume that it is desired to incorporate an integral-controller in the servo whose transfer-function is curve A of figure 41. The transfer-locus of the servo with incorporated integral-control is found by forming the product of the transfer-function of the controller with that of the servo. This operation is easily performed graphically. For example, if OB is the transfer vector of the servo without integral control for a particular frequency ω_B, and OB_i is the transfer vector for the integral-controller for this same frequency, the transfer-vector of the complete servo is found by taking the vector product of OB and OB_i. Thus, if OB_T is that vector product, the magnitude of OB_T is the product of the magnitudes of OB and OB_i, while the phase of OB_T is the sum of the phases of vectors OB and OB_i. The transfer locus of the integral-control servo is obtained by performing this vector multiplication for a range of frequencies and connecting with a smooth curve the tips of the vectors so located. If the transfer-loci of the components of the complete servo are plotted on suitable polar graph paper, the calculations involved in obtaining the locus of the cascaded system are minimized.

The transfer-function of the integral-controller with a particular value of time constant is plotted as curve B in figure 41. The construction of the transfer-locus of the integral-controller according to Equation (141) is illustrated in figure 42. The transfer vector of the

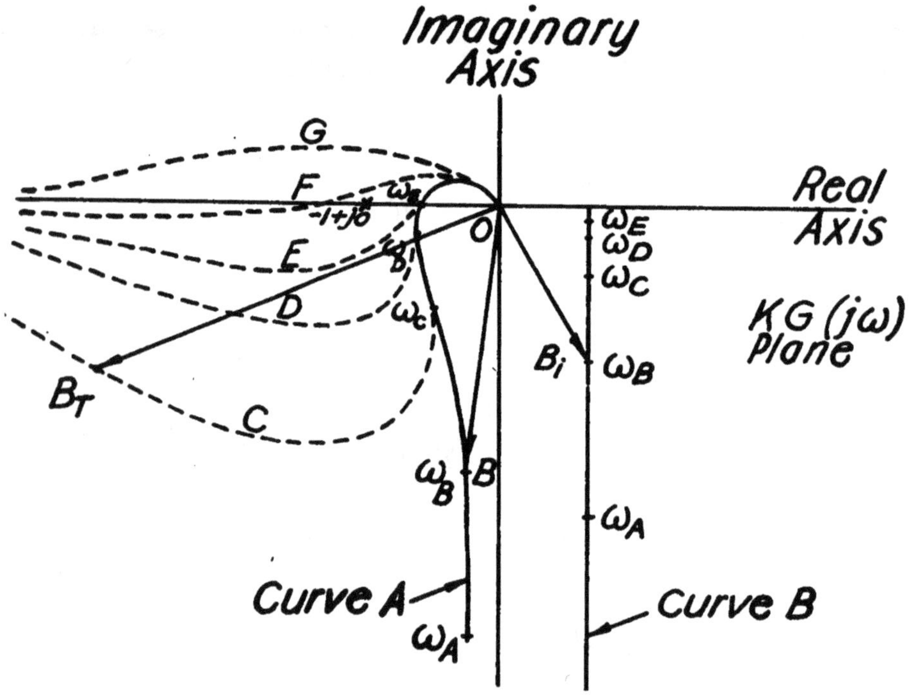

FIGURE 41

controller for a particular frequency ω_A is the sum of a fixed vector of unit length and an orthogonal vector whose length is inversely proportional to the product $RC\omega_A$. Therefore, changes

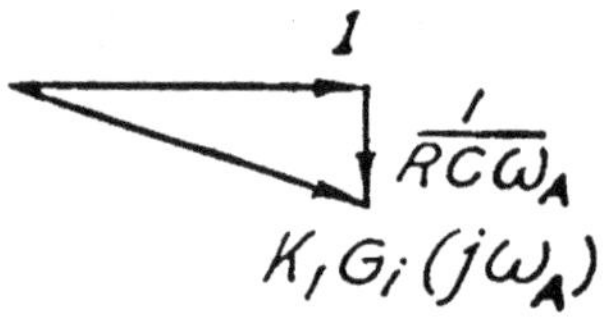

FIGURE 42

in the product RC do not affect the shape of the controller locus but only the value of the frequency corresponding to a particular point on the locus.

The effect of the integral-controller on the locus of the complete servo is pronounced at very low frequencies and negligible at high frequencies. The problem is so to choose the RC product that the locus is not shifted appreciably closer to the point (-1 + j0) since such a shift is an indication of a loss in the damping of the response. On the other hand, too large a value of RC product should not be selected or else the time required for the controller to correct a velocity-error is excessive.

The loci of the transfer-function of the complete servo for various values of the RC product are illustrated by figure 41, curves C, D, E, F, G, in the order of decreasing RC product.

The locus, curve C, corresponds to a controller with so large an RC product that appreciable shift in the locus, curve A, occurs only at very low frequencies. Curve D corresponds to a reduced time constant, but the reduction is not sufficient to affect the stability of the system, as shown by the fact that curve D does not approach too closely the point (-1 + j0). Locus E corresponds to a further reduction of the time constant. Although the system is still stable, its damping is poor as indicated by the fact that the maximum value of $\left|\frac{\theta_o}{\theta_i}(j\omega)\right|$ for locus E is approximately 3.5. When the time constant is reduced again, locus F results, corresponding to an unstable system. However, if the gain factor, K_1 , of the system is reduced, the intersection of curve F with the real axis can be made to occur on the right side of the point (-1 + j0) and the servo will be stable, although the damping of such a system is extremely poor. Finally, if the time constant, RC, is reduced to a sufficiently small value, the locus will not cross the negative real axis, as shown by curve G, and the system is unstable no matter how small the gain. The time constant corresponding to the locus of curve D is approximately the optimum for this example.

The loci of figure 41 reveal an interesting fact: In the case of a servo with an integral-controller it is not always possible to limit the maximum value of $\left|\frac{\theta_e}{\theta_i}(j\omega)\right|$ to a prescribed value by adjusting only the gain factor of the system. If the transfer-locus lies close to the point (-1 + j0), as in the case of locus E (or locus D), of figure 41, either decreasing or increasing the gain factor from its optimum value may increase the maximum value of $\left|\frac{\theta_o}{\theta_i}(j\omega)\right|$. If the damping is inadequate when the gain is set at its best value, altering the gain is of no assistance. This is in contrast with servo systems whose loci approach the negative imaginary axis at low frequencies, and in which the damping of the complex root always may be correctly chosen by adjusting the gain.

The integral-controller has been discussed in the preceding section as if it were an electronic active circuit. However, the principles and development are identical for mechanical, hydraulic, or other types of systems.

<u>Example: Application of Integral-Control to Servo with Third-Order Transfer-Function</u>

The principles involved in applying integral control to a servo become clearer if an example is analyzed. Therefore, consider the case in which an integral-controller is cascaded with the third-order transfer-function servo discussed in the preceding chapter. As explained there, the third-order transfer-function servo is idealized in certain respects, but it is sufficiently similar to physical servomechanisms that the conclusions resulting from its study are directly applicable in many instances. A block diagram of the components of the system is shown in figure 43.

The procedure that is followed in adjusting the parameters of the integral-controller and the control motor in order to obtain optimum servo performance is the same as that followed in the simpler case of the third-order servo with proportional controller. The $G(j\omega)$ loci of the system are plotted for a range of values of the adjustable parameters, and choice of a system is

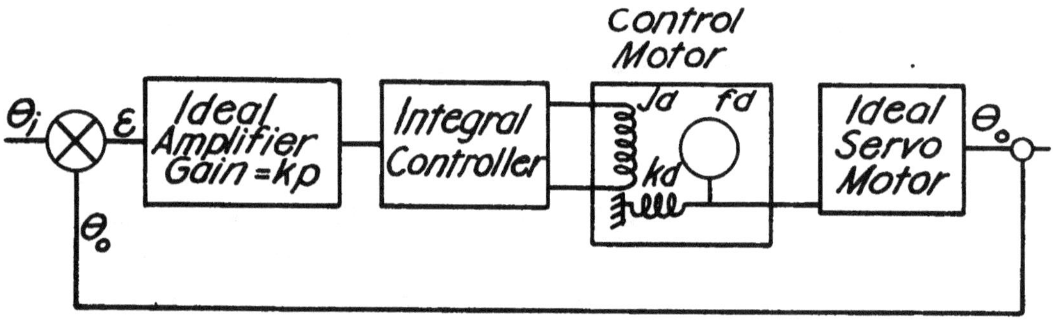

FIGURE 43

made upon the basis of the factors enumerated on page 45 - 46, with the added criterion that the RC product of the integral-controller should be as small as is compatible with well-damped servo operation.

The $G(j\omega)$ function of the third-order servo is

$$G_3(ju) = \frac{1}{ju(-u^2 + 2j\zeta_3 u + 1)}\ , \tag{123}$$

in which

$$u = \frac{\omega}{\omega_0}\ , \tag{119}$$

and ω_0 is the reference frequency defined by Equation (113). The $G(j\omega)$ function of the integral-controller is given by Equation (142):

$$G_1(j\omega) = 1 + \frac{1}{jRC\omega} \tag{142}$$

The function of Equation (142) should be transformed by making use of the same reference frequency, ω_0, used as a frequency base in the analysis of the third-order transfer-function. Applying transformation (119) to Equation (142), therefore, Equation (143) is obtained.

$$G_1(ju) = 1 + \frac{1}{jRC\omega_0 u} \tag{143}$$

$$\text{Let } RC\omega_0 = R_{13}\ , \tag{144}$$

in which R_{13} is a transformed time constant. Then,

$$G_1(ju) = 1 + \frac{1}{jR_{13}u}\ . \tag{145}$$

In the subsequent work the optimum value of the transformed time constant R_{13} is found. When an actual design of a physical system is carried through, the reference frequency, ω_0 , is known, and RC, the physical time constant of the integral-controller is calculated by means of Equation (144). The value of generalizing the development by means of a transformation such as

(119) has been discussed in the preceding chapter.

The $G(j\omega)$ function, $G_T(ju)$, of the complete servo including the integral-controller, is obtained by forming the product of Equations (123) and (145).

$$G_T(ju) = \frac{1}{ju(-u^2 + 2j\zeta_3 u + 1)} \left(1 + \frac{1}{jR_{13}u}\right) \tag{146}$$

The gain factor of the complete servo is:

$$K_T = K_1 K_3 = \frac{k_p k_m}{\omega_0 k_d}, \tag{147}$$

in which

K_T = the gain factor of the complete system;

K_1 = the gain factor of the integral-controller, equal to one for this controller;

K_3 = the gain factor of the third-order transfer-function servo, expressed in terms of the constants of the servo by Equation (122).

The transfer-function of the complete servo, $K_T G_T(ju)$, is obtained by forming the product of Equations (146) and (147).

$$K_T G_T(ju) = \frac{k_p k_m}{\omega_0 k_d} \frac{1}{ju(-u^2 + 2j\zeta_3 u + 1)} \left(1 + \frac{1}{jR_{13}u}\right) \tag{148}$$

The problem to be solved is either the proper choice of the transformed time constant R_{13} for a particular servo in which the damping ratio ζ_3 is fixed, or the selection of both R_{13} and ζ_3 if the damping ratio ζ_3 is adjustable. In both cases the proper system gain, K_T, must be determined.

The locus of Equation (146) has a general form illustrated by one of the curves C to G of figure 41. The intersection of the locus with the real axis is of considerable significance, since it is a measure of the maximum gain permitted by stable operation of the servo. The intercept is found to be

$$G_{T1} = \frac{1}{2\zeta_3 \left(\frac{2\zeta_3}{R_{13}} - 1\right)}. \tag{149}$$

If $\frac{2\zeta_3}{R_{13}} \geq 1$, G_{T1} is infinite or positive and no intersection occurs at any finite portion of the negative real axis. For stable operation the gain of the system must be so adjusted that the intersection lies to the right of the point $(-1 + j0)$: therefore, it is evident that unless

$$\frac{2\zeta_3}{R_{13}} < 1 \tag{150}$$

the servo is unstable no matter how small the gain. If condition (150) is met, stable operation may always be obtained with proper choice of the gain. Practically, however, $\frac{2\zeta_3}{R_{13}}$ must be much less than unity if a well-damped system is to result. Since K_T, the gain of the system, must be such that the unity point lies to the left of the intersection G_{T1} , the following inequality must exist if the servo is to be stable.

$$\frac{K_T}{2\zeta_3\left(\frac{2\zeta_3}{R_{13}} - 1\right)} < 1 \tag{151}$$

$$K_T < 2\zeta_3\left(\frac{2\zeta_3}{R_{13}} - 1\right) \tag{151a}$$

Inequalities(150) and (151) are the stability criteria of the third-order transfer-function servo with integral-controller.

Two guides are helpful in selecting the approximate values of damping ratio, ζ_3 , and the transformed time constant, R_{13} , previous to the actual plotting of the $G(j\omega)$ loci. First, the insertion of the integral-controller into the system should have a minimum effect upon the stability and damping of the servo. This means that the transfer-locus of the basic servo should remain approximately unaffected in the region of the point $(-1 + j0)$; and therefore, at frequencies corresponding to this region the transfer-function of the controller should have attained substantially its ultimate value of unit magnitude and zero phase-shift. Figure 42 illustrates the fact that the transfer-vector of the integral-controller is composed of a vector of fixed magnitude and phase, and a second vector, orthogonal to the first, whose magnitude is a function of the frequency, ω, and the RC product. The problem is to so choose the integral time constant that for frequencies larger than a prescribed value, the orthogonal vector is sufficiently small that the phase and magnitude of the composite vector is practically equal to that of the fixed vector. The geometry of the construction of figure 42 shows that the magnitude of the vector $K_1G_1(j\omega)$ will approach its final value more rapidly than the phase of the vector will approach zero. Thus

$$\left.\begin{aligned} &\text{If} \quad R_{13}u \gg 1 \\ &\left|G_1(ju)\right| \cong 1 \\ &\text{arc}\left[G_1(ju)\right] = \tan^{-1}\frac{1}{R_{13}u} \end{aligned}\right\} \tag{152}$$

Therefore, if the transfer-function of the integral-controller is not to affect the transfer-locus of the rest of the system at a particular frequency, it is necessary to so adjust the transformed time constant, R_{13} , that the phase-shift introduced at this frequency is less than the maximum that can be tolerated and the amplitude change will automatically be negligible.

The $G_3(ju)$ loci of the third-order transfer-function servo in figure 30 show that the maximum value of $\left|\frac{\theta_o(ju)}{\theta_i}\right|$ occurs at $u \cong 0.7$. At this frequency, therefore, the phase-angle of the transfer vector of the integral-controller should be small. A study of figure 30 leads to the estimate that if the phase-angle of $G_1(ju)$ is not more than about ten degrees at $u = 0.7$, the damping of the system should be unaffected by the insertion of the integral-controller. Therefore, relations (152) yield an approximately correct value of time constant, R_{13}, equal to eight.

The second design guide is based upon the fact that optimum servo adjustment results in a transfer-locus that tends to cross the real axis at right angles to it. The example of the preceding chapter proved that if the damping ratio, ζ_3 , was excessively large, the transfer-locus intersected the real axis at an acute angle, with a limiting effect on the gain, K_3, that could be used with the system. The effect of the integral-controller on the locus of the complete system is to make more acute the angle of intersection. This tendency can be counteracted somewhat by choosing a lower damping ratio, ζ_3 , for the combined system than that which has been chosen for optimum operation of the third-order transfer-function servo without the integral-controller. From such considerations it can be estimated that a damping ratio ζ_3 , equal to 0.3 to 0.5 will be optimum.

The loci of figures 44 and 45 were prepared with the above guides as starting points. Damping ratios of 0.3 and 0.5 and time constants of 2, 4, and 8 are the values of the parameters. The curve for which $h = 1.0$, figure 47, is a plot of the $G(ju)$ function in which $\zeta_3 = 0.4$ and $R_{13} = 6.0$. The shape of the curves verify closely the preliminary prediction of the optimum servo parameters. The conclusions are summarized below.

For the third-order transfer-function servo combined with the integral-controller, optimum servo parameters are:

$\zeta_3 = 0.4$ (approximately)

$R_{13} = 6$ to 8

$K_T = 0.3$ to 0.4

ζ_3 is defined by Equation (114a)

R_{13} is defined by Equation (144)

K_T is defined by Equation (147)

ω_0 is defined by Equation (113)

The above parameter values are the solution to the general problem in which all parameters are adjustable. If the damping ratio, ζ_3 , is fixed, the same general procedure is followed for the given value of ζ_3.

<u>Minimum Velocity-Error Systems</u>

It has been shown that the condition for a zero velocity-error system is that the gain of the controller be infinite at zero frequency, and that such a controller could be physically realized by a certain type of feedback amplifier provided the gain of the feedback path and that of the direct path were unity. This condition is met by requiring that the term "h" in Equation (138a) equal unity. If this requirement is met, the velocity-error of the servo is zero. The condition that h equal unity is a mathematical one, however, and while it may be approached as

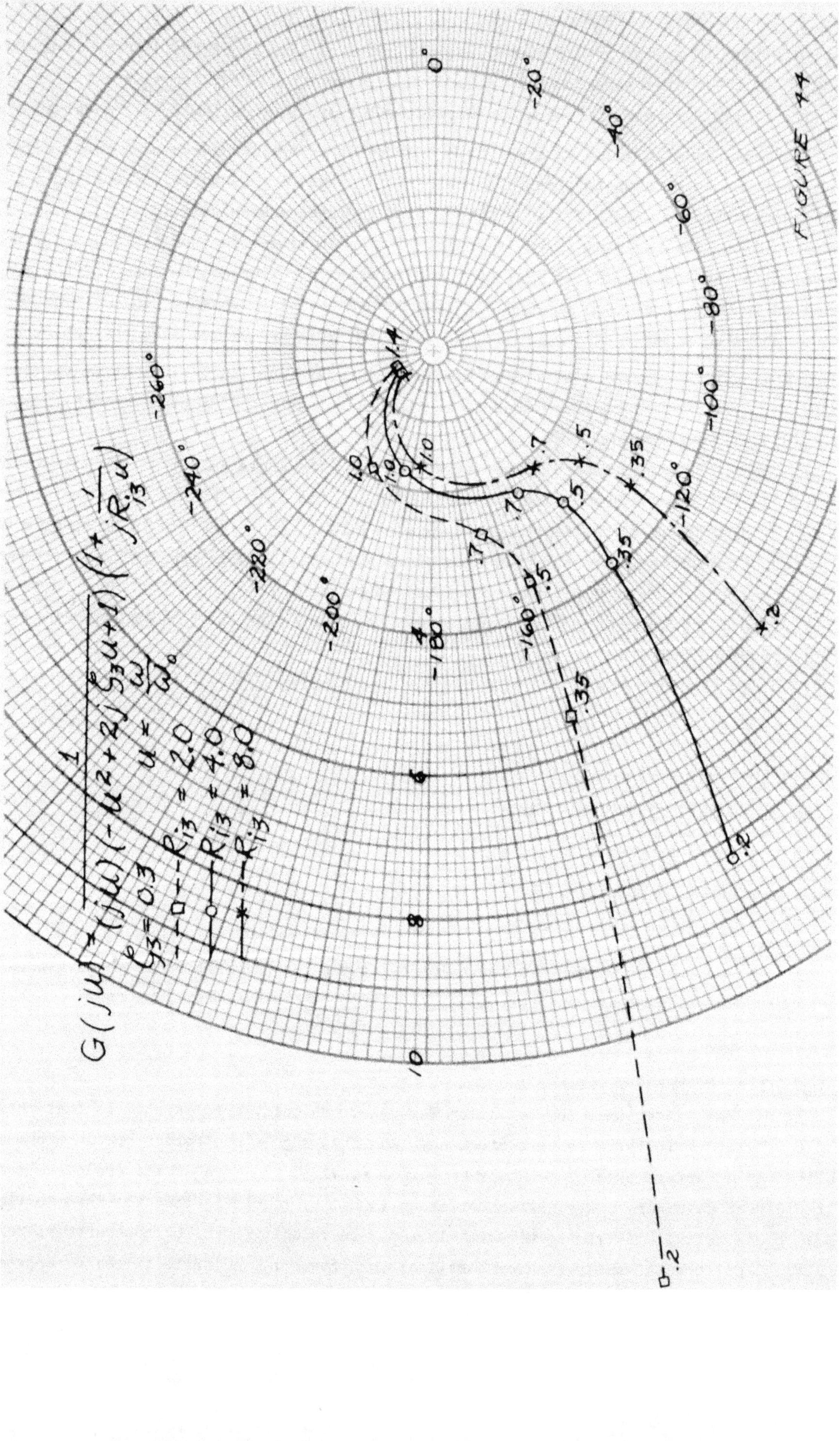

G(jω) = 1 / (jω)(−u² + 2j ξ₃u + 1) (1 + 1/(jR₁₃u))
ξ₃ = 0.3   u = ω/ω₀
R₁₃ = 2.0
R₁₃ = 4.0
R₁₃ = 8.0
0°
−20°
−40°
−60°
−80°
−100°
−120°
−160°
−180°
−200°
−220°
−240°
−260°
FIGURE 44

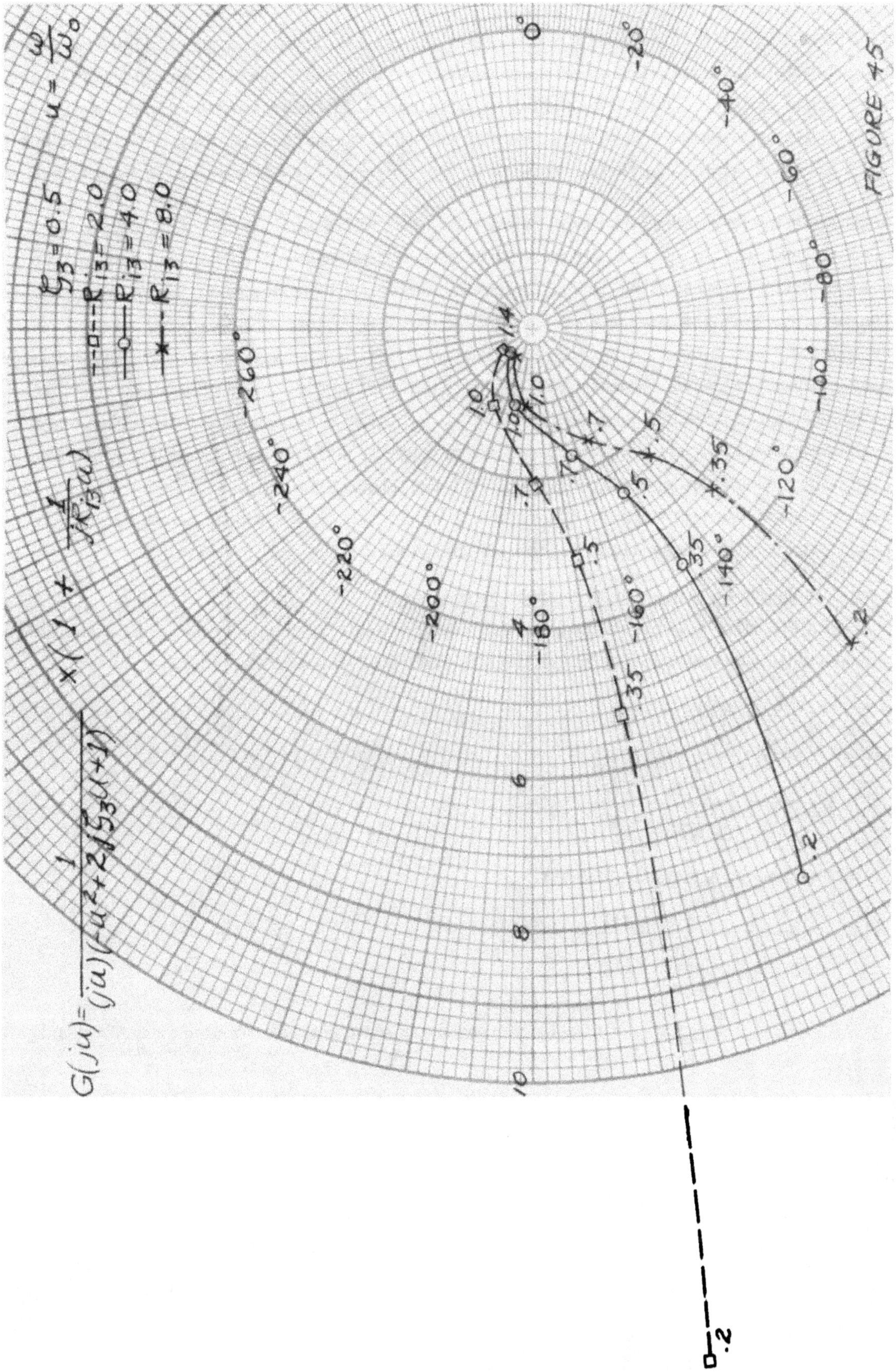

FIGURE 45

closely as desirable, it is impossible to design a physical system in which this relation is achieved and maintained with mathematical exactness. Therefore, both cases in which the factor, h, is larger and smaller than unity should be considered in order to determine the effect of small variations in h above and below unity. The velocity-error of a system in which h = 1 is not zero; it is diminished, however, by the factor ($\frac{1 - h}{h}$) from the value it would have were the integral-controller absent and may, therefore, be sufficiently small for practical purposes. Such servo systems can be called minimum-velocity-error systems; the integral-controller is designated as an under-compensating integral-controller or an over-compensating integral-controller depending upon whether the parameter h is smaller or larger than unity.

The velocity-error of most servomechanisms need not be absolutely zero for satisfactory performance since the servo is never subjected to an input which is purely a constant velocity. The input may be predominantly constant velocity, but will always contain disturbances due to both changes in the true input function and false signals arising from gear irregularities, imperfections in the data transmission system, and the like. The system is generally considered satisfactory if its velocity-error is less than, or of the same order of magnitude as, the transient errors produced by these disturbances in the input.

Case I: $h < 1$ Under-Compensating Integral-Controller

The first and more important case of the minimum velocity-error system to be considered is that in which h is less than unity and Equation (138a) of the controller is

$$K_i G_i(s) = \frac{(1 + RCs)}{1 - h + RCs} \tag{138a}$$

In the above,

$$\left.\begin{aligned} G_i(s) &= \frac{1}{1 - h} \frac{1 + RCs}{1 + \frac{RC}{1 - h} s} \\ K_i &= 1 \end{aligned}\right\} \tag{138b}$$

In terms of frequency, Equation (138b) is

$$G_i(j\omega) = \frac{1}{1 - h} \frac{1 + RC\,(j\omega)}{1 + \frac{RC}{1 - h}\,(j\omega)} \tag{153}$$

When the expression for $G_i(j\omega)$ is non-dimensionalized with respect to frequency by means of the transformation

$$\frac{\omega}{\omega_0} = u, \tag{119}$$

Equation (119) becomes

$$G_i(ju) = \frac{1}{1 - h} \frac{1 + RC\omega_0(ju)}{1 + \frac{RC\omega_0}{1 - h}\,(ju)} \tag{154}$$

The locus of Equation (154) is a semi-circle located in the fourth quadrant of the complex plane with a center on the positive real axis at $\frac{(2 - h)}{2(1 - h)}$ and a radius equal to $\frac{h}{2(1-h)}$. The locus is illustrated in figure 46. The vertical distance, ac = $\frac{h}{u_1 R\omega_0 C}$ is laid off from the intercept nearer the origin and a line, bc, is drawn to the outer intercept. The intersection of the line, bc, with the circle at d is the tip of the vector defined by Equation (154) for the frequency u_1.

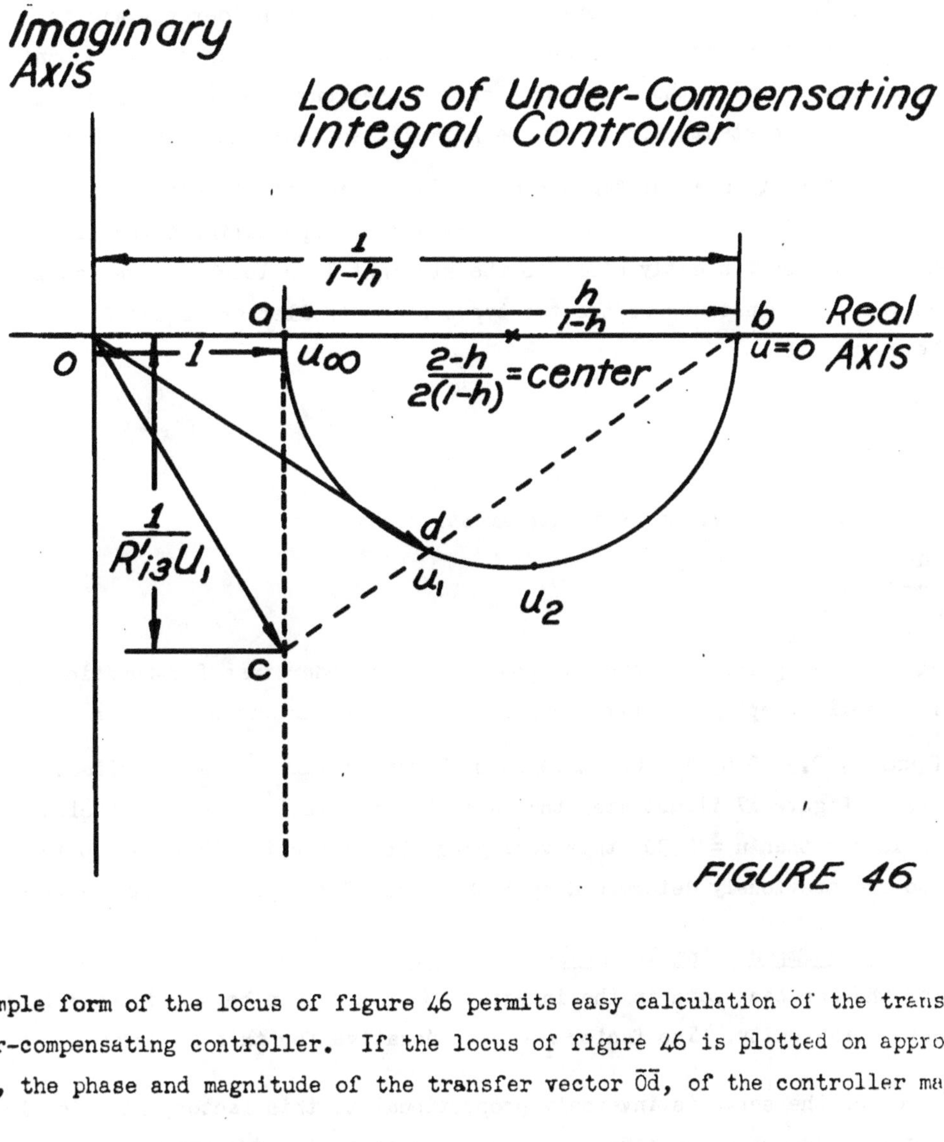

FIGURE 46

The simple form of the locus of figure 46 permits easy calculation of the transfer-function of the under-compensating controller. If the locus of figure 46 is plotted on appropriate polar graph paper, the phase and magnitude of the transfer vector $\overline{Od}$, of the controller may be read at once.

Let the following substitution be made

$$R_{13}' = \frac{RC\omega_0}{h} \tag{155}$$

in which R_{13}' is the transformed time constant of the under-compensating integral-controller. Then, if R_{13}' is equal to R_{13}, the time constant of the controller when h = 1, the vector $\overline{Oc}$,

figure 46, is the transfer vector of an ideal integral-controller for the same frequency, u_1.
The comparison between the two vectors $\overline{Oc}$ and $\overline{Od}$ is readily made. The difference is small for frequencies at which the vertical line, ac, and the circle approach tangency. At larger frequencies the difference between the two vectors is primarily a phase difference; appreciable differences in magnitude do not occur until positions in the vicinity of u_2 have been reached by the transfer-vector of the under-compensating controller.

The presentation of the transfer-functions of the ideal integral- and under-compensating integral-controllers in the graphical form of figure 46 is of considerable aid to the visualization of the design problem and enables design criteria to be obtained with a manimum of actual calculation. For example, the discussion of the previous section on integral-controllers has shown that the time constant, R_{13}, must be so adjusted that the phase-shift for certain frequencies is not larger than about ten degrees. Inspection of circular loci show that at this value of phase-shift the circle diagram of the under-compensating integral-controller for values of $(1-h) < 0.3$ is sufficiently close to the straight line locus of the ideal integral-controller that the time constants of the two types of controller are equal for similar servo operation. Therefore, if

$$(1 - h) < 0.3$$

$$R_{13}' \cong R_{13} \tag{156}$$

The actual RC product in the two cases is different, since

$$RC = \frac{R_{13}'h}{\omega_0} \tag{155a}$$

The effect of varying the parameter h upon the third-order transfer-function servo with integral-controller while keeping the time constant, R_{13}', invariant is illustrated in figure 47. Values of h of unity, 0.9, 0.8, and 0.7 have been chosen to illustrate the effect of variations in this parameter. Figure 47 illustrates the fact that the loci are not appreciably different for frequencies larger than u = 0.30, thus verifying the conclusion that the optimum value of R_{13}' is equal to the previously determined optimum value of R_{13} provided that $0.7 < h < 1.0$.

Case II: $h > 1$: Over-Compensating Integral-Controller

The gain of the feedback path in the integral-controller can be so adjusted that the parameter h is greater than unity. The factor $\frac{1}{1 - h}$ is negative for this adjustment, and since the the velocity-error of the servo is inversely proportional to this factor, it, too, is negative. The physical significance of a negative velocity-error is that under constant velocity input conditions the servo output leads the input instead of lagging it as it would do were h less than unity. This in itself may or may not be a desirable result, but in any case, it is necessary to determine the effect upon the servo stability of an over-compensating integral-controller in order that the effect of variations in the circuit components caused by aging, wear, temperature, etc., may be predicted.

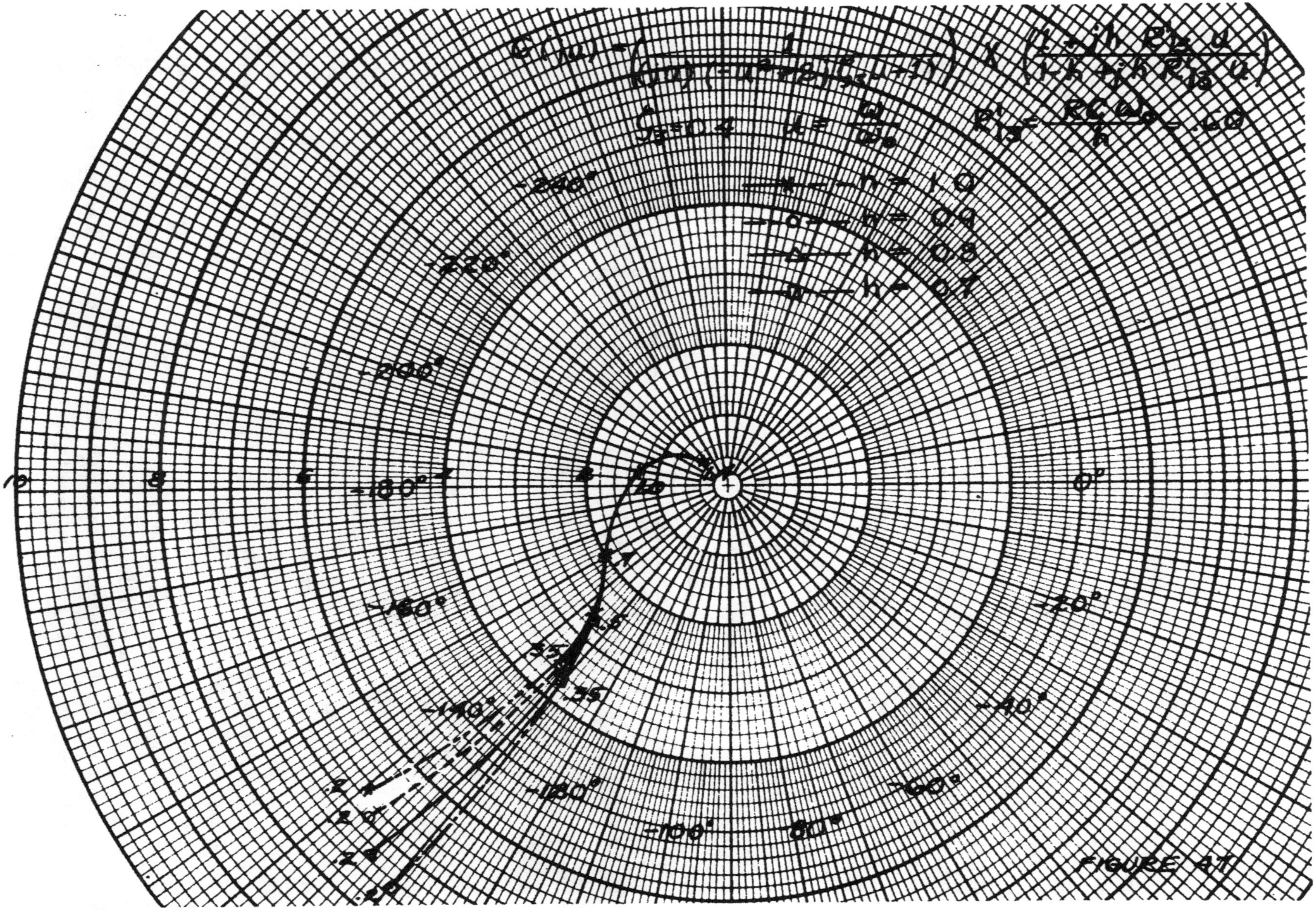
h = 1.0
h = 0.9
h = 0.8
h = 0.7
0°
-20°
-40°
-60°
-80°
-100°
-120°
-140°
-160°
-180°
-200°
-220°
-240°
2
4
6
8
10

FIGURE 47

The locus of the transfer-function of the over-compensating integral-controller is illustrated in figure 48. The locus is a circle similar to that of the under-compensating controller shown in figure 46 except that the zero frequency intercept $\frac{1}{1 - h}$ is located on the negative real axis for the case of the over-compensating controller and, therefore, the circle locus is shifted to a somewhat different position in the complex plane. The construction for determining the position of the transfer-function vector is exactly the same as that for Case I, in which $h < 1$. The most important effect of this variation in circle position is that the phase of the transfer-function vector no longer varies from zero to a maximum and back to zero, but varies unidirectionally from zero to 180° as the frequency decreases from infinity to zero. The effect of this type of transfer-function on the transfer-locus of a servomechanism is illustrated in figure 49 in which curve A is the transfer-locus of a typical servo, curve B is the locus of the over-compensating integral-controller, and curve C is the locus of the combination. The transfer-locus of the combined system crosses the negative real axis twice, and the gain must be so

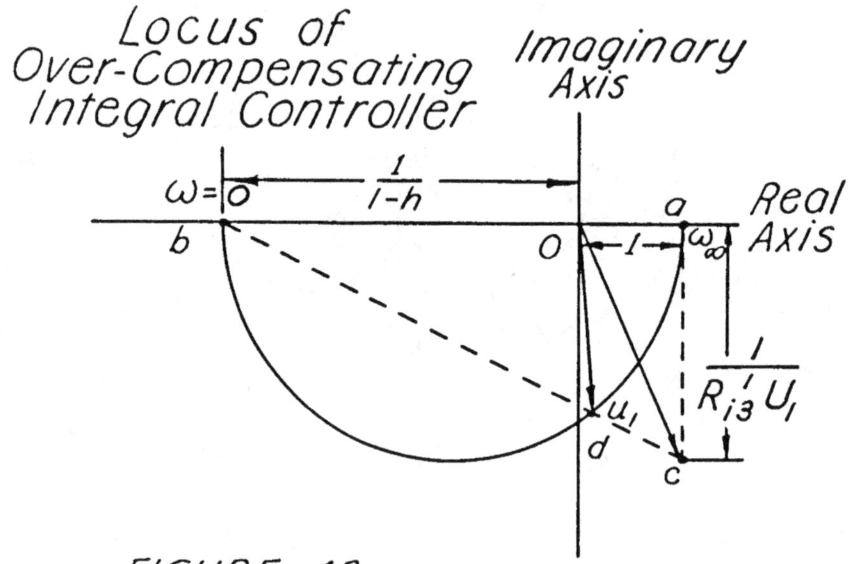

FIGURE 48

adjusted that the unity point lies between the intersections. Since such a system may be unstable if the gain is either too high or too low, the controller must be so designed that its time constant is not only sufficiently large that the shift introduced in the region of the inner intercept is small (the design criterion if $h = 1$) but also that the outer intercept is distant from the unity point. Fortumately, these two criteria are easily satisfied for values of h near unity, which are the values that are most often met in practice. A large value of h, however, decreases the separation between the inner and outer intercepts and thus exerts considerable unstabilizing influence. The outer intercept always can be increased in value by increasing the time constant of the integral-controller over that which is set by the phase-shift requirements at the inner intercept, but this procedure results in an increase in time required for the servo to reach its minimum error when a constant velocity is suddenly applied to the input.

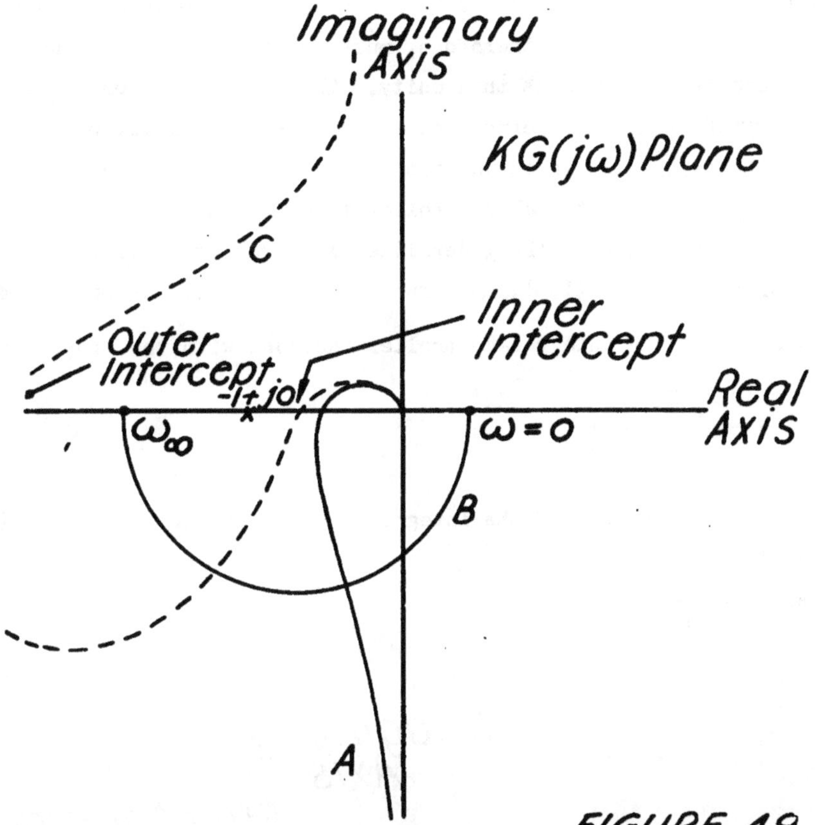

FIGURE 49

Figure 50 illustrates the effect upon the transfer-vector of the integral-controller of varying the term, h, while maintaining the time constant R_{13}' at a constant value. Vector $\overline{Oa}$ is the transfer-vector of an under-compensating controller in which h is set at 0.9. Vector $\overline{Ob}$ is the corresponding vector for the ideal integral-controller in which h equals unity, and vectors $\overline{Oc}$, $\overline{Od}$, and $\overline{Oe}$ are the corresponding transfer-vectors of the overcompensating controller for which h is set at 1.1, 1.2, and 1.3, respectively. The fact that vectors $\overline{Oa}$ to $\overline{Oe}$ correspond respectively to progressively more unstable controllers is shown by the increased phase-angle each vector possesses.

The third-order transfer-function servo is taken as an example for illustrating the effect of overcompensating integral control upon a servomechanism. The time constant, R_{13}' , is equal to 6.0 and values of h are 1.1, 1.2, and 1.3. The loci are plotted in figure 51. The actual RC product of the controller is calculated from Equation (155a). The results of figure 51 have the expected form. Larger values of h bring the low frequency end of the curve closer to the real axis and while the scale used prevents the intercepts from being plotted, the form the curve takes for very low frequencies is not essentially different from that illustrated in figure 49. The general conclusion may be drawn that the parameter h is best kept as near as poss ble to unity, but there is little loss in servo stability for values of h as large as 1.2 or 1.3.

High Gain Integral-Controller

The integral-controller that has been considered throughout this chapter is one in which the gain factor is unity. It, therefore, must be cascaded with an amplifier for all applications which require higher gains than unity. Thus, when the controller is cascaded with the third-order transfer-function servo, the proportional amplifier of gain, k_p, is retained, as indicated in figure 43. It is advantageous in a physical design, however, to combine the proportional amplifier and the integral-controller into one unit.

The transfer-function of the generalized integral-controller (ideal, over or undercompensating) is given by Equation (133); the transfer-function of the proportional amplifier is k_p; the transfer-function of the integral-controller cascaded with the proportional amplifier is the product

$$K_p G_i(s) = \frac{k_p H(s)}{1 - hH(s)L(s)} \tag{157}$$

In the preceding development of the integral-controller the function, H(s), has been set equal to unity, yielding

$$K_p G_i(s) = \frac{k_p}{1 - hL(s)} \tag{158}$$

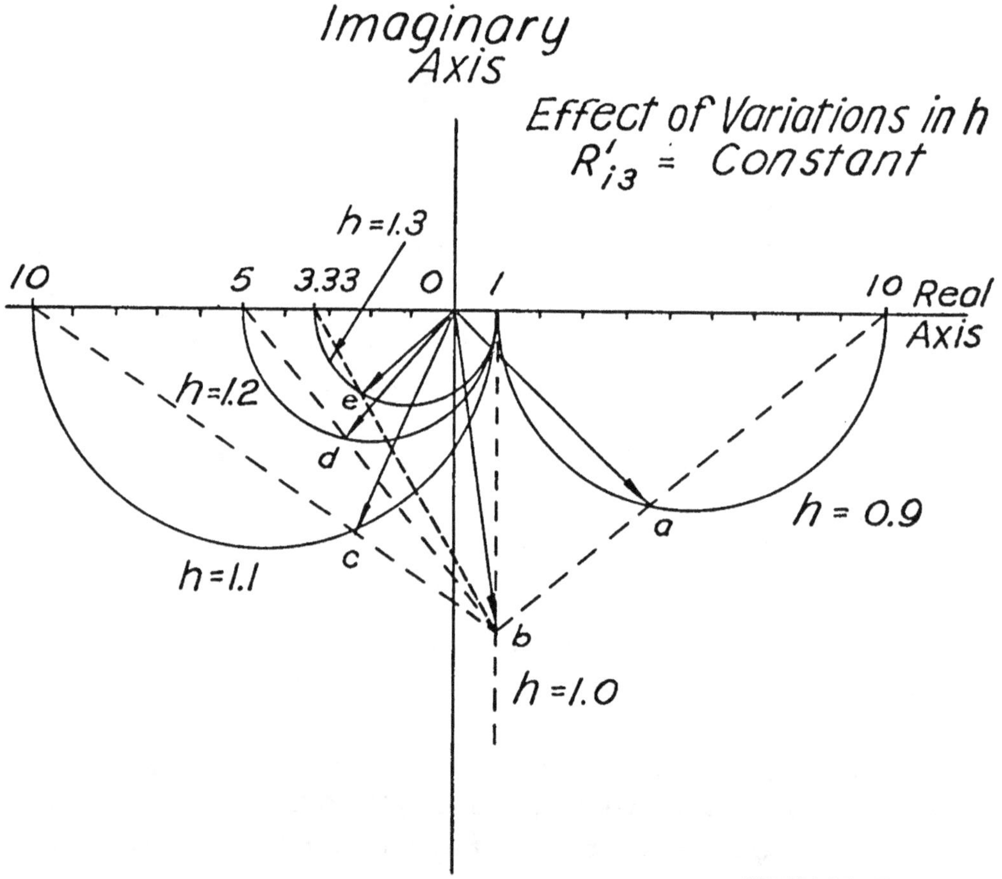

FIGURE 50

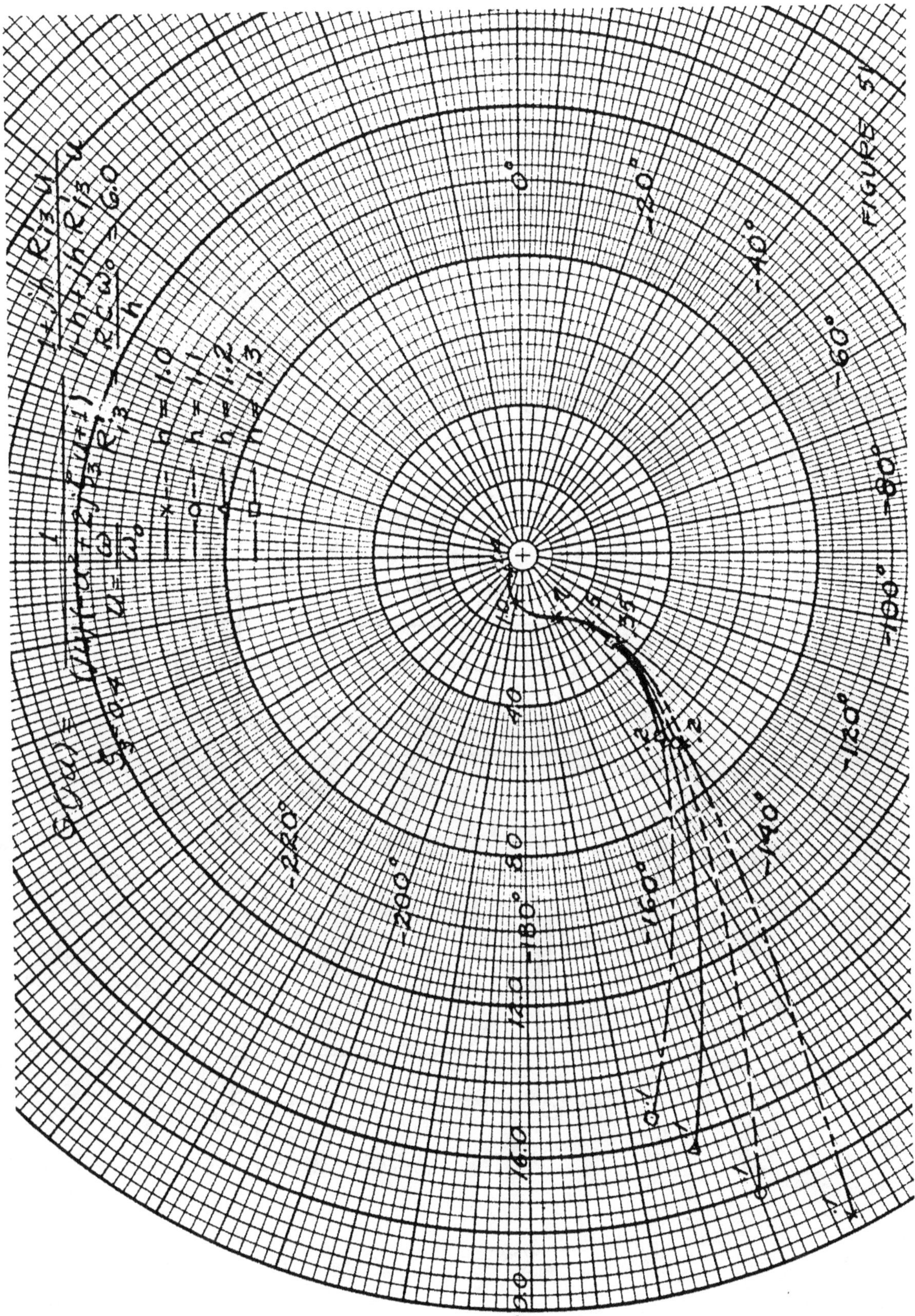
0°
-20°
-40°
-60°
-80°
-100°
-120°
-140°
-160°
-180°
-200°
-220°
4.0
8.0
12.0
16.0
20.0
h = 1.0
h = 1.1
h = 1.2
h = 1.3

FIGURE 51

In place of a direct path gain, H(s), equal to unity, let the direct gain equal k_p.

$$H(s) = k_p \tag{159}$$

and instead of feeding h times the output voltage back into the feedback path, let $\frac{h}{k_p}$ times the output voltage be applied to the feedback link. The transfer-function of the system thus altered is:

$$K_i G_i(s) = \frac{k_p}{1 - hL(s)} \tag{160}$$

Since expressions (160) and (158) are identical, the altered integral-controller has a response identical to that of the previous integral-controller cascaded with the proportional amplifier. The block diagram of the high gain integral-controller is figure 52.

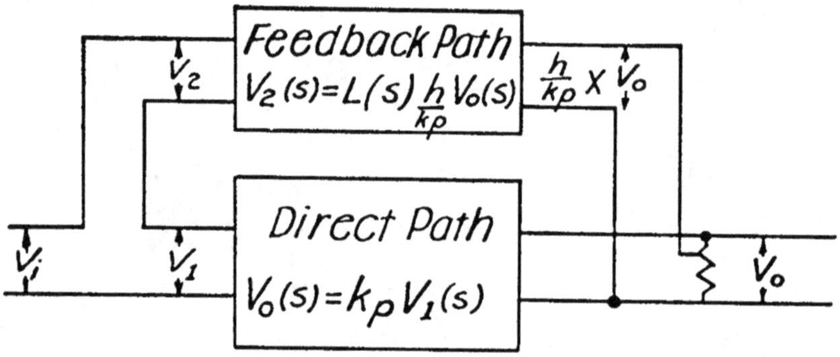

FIGURE 52

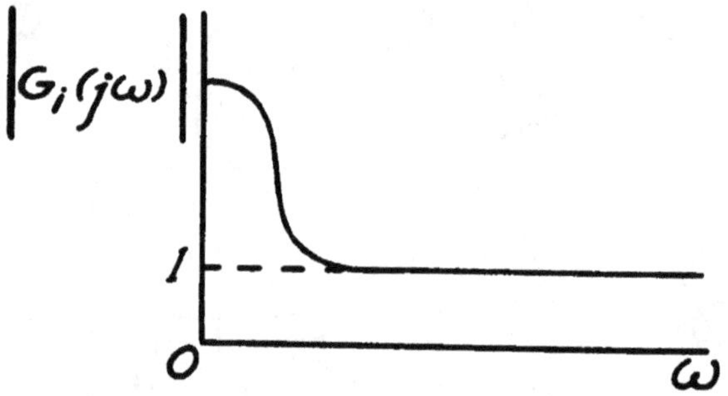

FIGURE 53

Realization of Undereompensating Integral Control by Cascade Circuits

The previous sections have indicated how ideal integral control can be obtained by means of an amplifier employing a feedback path so adjusted that the gain of the amplifier is infinite at zero frequency. Since ideal integral control requires infinite gain at zero frequency, and infinite gain can be obtained only by regenerative circuits, true integral control can be obtained by no means other than those employing regenerative devices. It can be shown that regenerative feedback amplifiers are often unsatisfactory becuase of an inherent tendency to drift, and that if this tendency cannot be overcome, this type of coptroller is excluded from use in certain applications.

It has also been shown that ideal integral control is often unnecessary and that undercompensating integral control is entirely adequate for most applications. Since this type of control does not require infinite gain, it is possible and highly desirable to develop circuits for its realization that do not employ regenerative devices.

The amplitude response of the undercompensating controller obtained from tha circle diagram of figure 46 is illustrated in figure 53. The response is large at very low frequencies, drops rapidly with frequency, and approaches unity asymptotically. Since the magnitude of the response at zero frequency is large but not infinite, the same characteristic might be obtained by cascading a filter for controlling the shape of the amplitude curve with an amplifier to give the correct gain level.

A simple filter with an amplitude characteristic of the general form of figure 53 is that illustrated in figure 54. The transfer-function of this network is

$$\frac{V_2}{V_1}(s) = \frac{1 + R_2C_1s}{1 + (R_1 + R_2)C_1s} \qquad (161)$$

in which R_1, R_2, and C are the resistances and capacitances shown in figure 54. Suppose this filter is cascaded with an amplifier whose gain is k_a; the transfer-function $K_aG_a(s)$ of the filter and the amplifier combined is

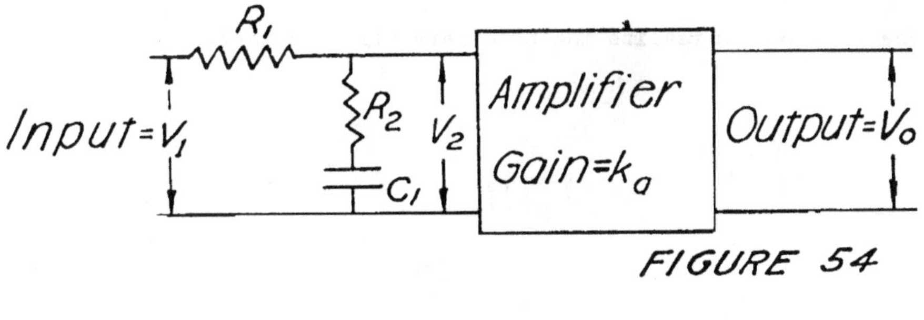

FIGURE 54

$$K_aG_a(s) = \frac{k_a(1 + R_2C_1s)}{1 + (R_1 + R_2)\ C_1s} \tag{162}$$

If Equation (162) is compared with the expression for the transfer-function of the undercompensating integral-controller, Equation (138b), it is seen that the two expressions are similar in form. The two expressions

$$G_1(s) = \frac{1}{1 - h}\ \frac{1 + RCs}{1 + \frac{RCs}{1 - h}} \tag{138b}$$

are identical and the two circuits will have identical characteristics provided that the gain of the amplifier and the circuit components of figure 54 are so adjusted that the following relations exist:

$$k_a = \frac{1}{1 - h} \tag{163}$$

$$\frac{R_1}{R_2} = \frac{h}{1 - h} \tag{164}$$

$$(R_1 + R_2)C_1 = \frac{RC}{1 - h} \tag{165}$$

Equations (163), (164), and (165) hold for all values of h <u>less</u> than unity.

The two circuits being equivalent have the same locus in the complex plane, and conclusions drawn for one apply to the other. An important difference exists, however, in the size of the physical components that are necessary to attain the proper time constant. Equation (165) shows that for equivalent servo performance the resistance-capacitance product of the cascade circuit must be larger by a factor of $\frac{1}{1 - h}$ than the RC product of the regenerative circuit. This fact, unsuspected from a casual investigation, is very significant; for certain applications it proves to be the limiting factor in the usefulness of the cascade circuit of figure 54, since for values of h close to unity the factor $\frac{1}{1 - h}$ becomes very large, and excessively large physical components may be required to realize the necessary time constant.

CHAPTER VI

THEORY OF PHASE-LEAD-CONTROLLERS

The preceding chapter has developed the theory of the design of devices that compensate for the steady-state error of a servomechanism. These devices are known as steady-state error compensators (integral-controllers) and are so referred to in the block diagram of figure 4. It has been pointed out previously that a servo system may be unacceptable if its transient errors are excessive, and one of the devices illustrated in block form in figure 4 is a device for compensating for that transient or dynamic error. The theory of the design of such devices is developed in this chapter.

It has long been known that improved servo performance is secured by employing as a servo-controller a device whose output is proportional not only to its input but also to a function approximating the time rate of change of its input. In certain instances further improvement is obtained if the output of the controller is proportional not only to the input and time rate of change of the input but also to one or more higher order time derivatives of the controller input. The operation of these servo-controllers, sometimes termed derivative controllers, is explained in a degree by reasoning heuristically that the derivative controller is able to "foretell" the future position of the servo input and therefore so control the servo motor that the servo output follows the input with reduced error. Reasoning on such an intuitive basis, however, is unable to produce a concrete servo design. In order to obtain a satisfactory analysis of such systems, the controller must be mathematically specified and a reasonably exact analysis carried out. In the past this has been done by solving the differential equation of the derivative-controller and servomechanism with which it is employed. Unfortunately, however, the complexity of the mathematics encountered if the order of the servo characteristic equation is above three has led many to make simplifying assumptions concerning the functional relationship of the controller and servo that are not justified and have resulted in erroneous conclusions. Such assumptions are unnecessary if the servo design is obtained by analyzing the transfer-loci of the system, and consequently, correct conclusions may be reached provided a proper procedure is followed.

This chapter develops the design of so-called derivative controllers by studying their transfer-loci together with those of the servo systems in which they are incorporated. Rigorous design methods are employed, and physical restrictions are placed upon the derivative controllers. It may be shown that an ideal derivative controller, that is, a device the output of which is proportional to the sum of the input and the time derivative of the input, cannot be physically realized. However, it is possible to approximate an ideal derivative controller by employing circuits (electrical, mechanical, etc.) whose output possesses such a phase characteristic that over a limited frequency range the slope of the phase vs. frequency function is positive. The requirement that the controller possess this positive phase-frequency characteristic is more fundamental than the requirement that it approximate an ideal derivative controller. Such controllers, therefore, are termed phase-lead-controllers, or simply lead-controllers, with the understanding that the name signifies a device with a positive rate of change of phase as a function of frequency. In the past, the difficulty with which higher order ideal derivative controllers (second-derivative controllers, third-derivative controllers, etc.,) are approximated has led many to believe that such devices are of little value in servomechanism synthesis. If physical devices of this nature are considered as devices whose phase vs. frequency response is positive over a certain frequency range, it can be shown that they have a very definite use in certain applications, and their true value may be more accurately appraised.

As an introduction to the subject of the design of phase-lead-controllers, an ideal derivative controller is briefly considered and difficulties pointed out which are encountered when such an ideal device is assumed to be realizable.

Derivative Controller

A common means of functionally specifying a servo controller is by writing the differential equation relating the controller output and input. Thus, if the output of a controller is $\theta_d(t)$, and the input to the controller is the servo error, $\varepsilon(t)$, (according to the convention illustrated by figure 55) the differential equation relating $\theta_d(t)$ and $\varepsilon(t)$ specifies the controller.

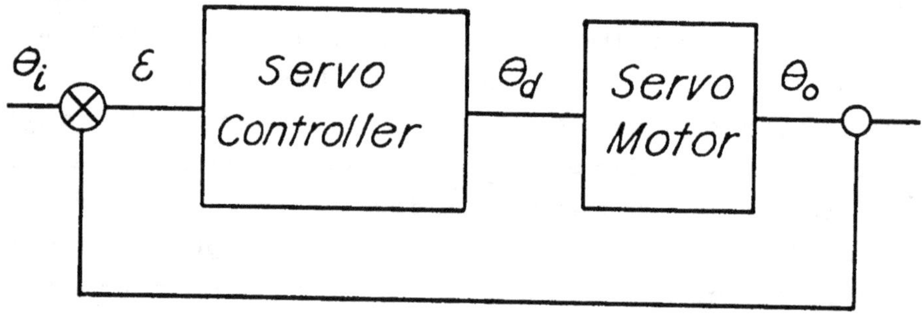

FIGURE 55

If the controller is a proportional device, the equation is

$$\theta_d(t) = K_p \varepsilon(t) \qquad (166)$$

in which K_p is a proportionality factor. If the controller is a device whose output is proportional to both its input and the time derivative of its input, it is represented by the following differential equation:

$$\theta_d(t) = K_d \varepsilon(t) + k_d K_d \frac{d}{dt} \varepsilon(t) \qquad (167)$$

Thus, the output of a derivative controller is composed of two parts: one that is related by a proportionality factor K_d to the controller input, and another that is related by the factor $k_d K_d$ to the derivative of the controller input. The reason for writing the controller function in this particular manner becomes evident in the following equation. The equivalent Laplacian equation of (167) is

$$\theta_d(s) = K_d \, (1 + k_d s) \, E(s) \qquad (168)$$

The transfer-function $K_dG_d(s)$ of the ideal derivative controller is

$$K_dG_d(s) = \frac{\theta_d(s)}{E\ (s)} = K_d(1 + k_ds) \tag{169}$$

in which the gain factor is equal to K_d and the frequency dependent portion $G_d(s)$, of the transfer-function is

$$G_d(s) = 1 + k_ds \tag{170}$$

The transfer-function of the controller as a function of frequency is obtained by replacing the complex variable, s, by the frequency, $j\omega$.

$$K_dG_d(j\omega) = K_d(1 + k_dj\omega) \tag{171}$$

The locus of (171) in the complex plane is illustrated by curve A of figure 56.

One explanation of the action of the derivative controller in improving the system response is as follows: Assume that the transfer-locus of a servo system with a proportional controller is curve B, figure 56. The transfer-locus of that system with a derivative controller is obtained by forming the vector product of the transfer-locus of the original system with that of the derivative controller. The transfer-locus of the complete system is a locus such as curve C of figure 56. It is seen that the effect of the derivative controller is to shift the locus of the original system in a positive angular direction. This results in an increase in the frequency at which $\left|\frac{\theta_{o(j\omega)}}{\theta_i}\right|$ is a maximum which in turn indicates that the natural frequency of the system is increased resulting in increased speed of response of the servo. The improvement is primarily due to the fact that the phase of the derivative controller increases positively with frequency and therefore, compensates for the increasing <u>negative</u> angle of the original servo system.

A servo system that has been employed to illustrate the effect of a derivative controller is that illustrated by figure 9, Chapter II. This system, termed a type I servo by Brown,[6] comprises a servo motor the output of which is characterized by a moment of inertia, J_L, and a viscous damping, f_L, and the torque of which is proportional to the servo error. The equations of the system are:

$$T(s) = k_pE(s) \tag{172}$$

(The torque, T, of the motor is proportional to error.)

$$T(s) = (J_Ls^2 + f_Ls)\ \theta_o(s) \tag{173}$$

(The equation of motion of the motor.)

The transfer-function, $K_1G_1(s)$, of this system is given by Equation (174).

$$K_1G_1(s) = \frac{\theta_o(s)}{E\ (s)} = \frac{k_p}{J_Ls^2 + f_Ls} \tag{174}$$

Let $\frac{J_L}{f_L}$ (175)

Equation (174) can then be written:

$$K_1G_1(s) = \frac{k_p}{f_L s\ (\tau_L s + 1)} \tag{176}$$

Since the moment of inertia, J_L, and the viscous damping, f_L, are determined primarily by the servo load or the design of the servo motor, they can be considered more or less fixed, and a frequency transformation of the type explained on Page 46 can be made. Therefore, let $\tau_L s = s'$ (177)

Substituting (177) into (176),

$$K_1G_1(s') = \frac{\tau_L k_p}{f_L s'\ (s' + 1)} \tag{178}$$

If (178) is grouped as follows, $K_1G_1(s') = \frac{\tau_L k_p}{f_L}\ \frac{1}{s'\ (s' + 1)}$ (178a)

it is evident that $K_1 = \frac{\tau_L k_p}{f_L}$ (179)

$$G_1(s') = \frac{1}{s'\ (s' + 1)} \tag{180}$$

In terms of the transformed frequency, $u = \omega\tau_L$,

Equation (178a) becomes

$$K_1G_1(ju) = \frac{\tau_L k_p}{f_L}\ \frac{1}{ju(1 + ju)} \tag{181}$$

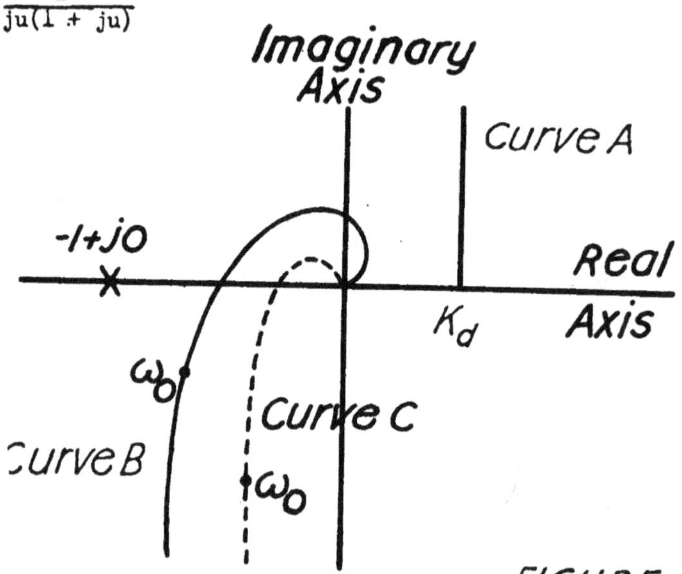

FIGURE 56

The locus of (181) is plotted in figure 57, curve A. The gain factor, K_1 , that limits the maximum value of $\left|\frac{\theta_o(ju)}{\theta_i}\right|$ to one and one-third, determined in the manner described in Chapter IV, is found to be 1.4. The resonant frequency, u_m, at which the maximum value of $\left|\frac{\theta_o(ju)}{\theta_i}\right|$ occurs is approximately equal to unity. A summary of the design constants of the type I system follows:

$$K_1 = \frac{\tau_L k_p}{f_L} = 1.4 \qquad (182)$$

$$\tau_L = \frac{J_L}{f_L} \qquad (175)$$

$$u_m = \tau_L \omega_m = 1.0 \qquad (183)$$

When the proportional controller of figure 9 is replaced by an ideal derivative controller, the transfer-function of the new system is obtained by substituting the transfer-function (169) of the derivative controller for k_p, the transfer-function of the proportional controller, in (176). The transfer-function $K_T G_T(s)$, of the new system is given by Equation (184).

$$K_T G_T(s) = K_d\ (1 + k_d s)\ \frac{1}{f_L s\ (\tau_L s + 1)} \qquad (184)$$

If the transformation of (177) is applied to (184), Equation (185) is obtained.

$$K_T G_T(s') = K_d \left[1 + \frac{k_d}{\tau_L}\ s' \right] \left[\frac{\tau_L}{f_L s'\ (s' + 1)} \right] \qquad (185)$$

In terms of the frequency $u = \tau_L\ \omega$, the transfer-function is

$$K_T G_T(ju) = \frac{K_d\ L}{f_L}\ \frac{1 + \frac{k_d}{L}\ ju}{ju\ (1 + ju)} \qquad (186)$$

The term k_d is the derivative constant and may be adjusted to equal τ_L, the time constant of the servo motor. If this adjustment is made, Equation (186) reduces to

$$K_T G_T(ju) = \frac{K_d \tau_L}{f_L}\ \frac{1}{ju} \qquad (187)$$

The function of Equation (187) is the transfer-function of the ideal zero displacement-error servo, as a comparison of (187) and Equation (93) reveals. The discussion beginning with page 42 has shown that the ideal zero displacement-error servo may be made as <u>fast as desired</u>, simply by increasing the gain factor of the system.

The transfer-loci of the basic servo system (the type I servo), the derivative controller, and the combined system are curves A, B, and C respectively of figure 57.

The preceding development has demonstrated the desirability of employing a derivative controller, but the development has led to a physically unrealizable result. The ideal zero displacement-error servo is physically unrealizable for a reason pointed out earlier: namely, the transfer-function of any physical system cannot be of lower order than two. Therefore, while the above derivation is convincing proof that a derivative controller augments servo performance, it has failed to shed light on the following questions that must be answered by a satisfactory design procedure: (1) what are the values of the system parameters that produce optimum servo performance? (2) what is that optimum performance? The first question must be

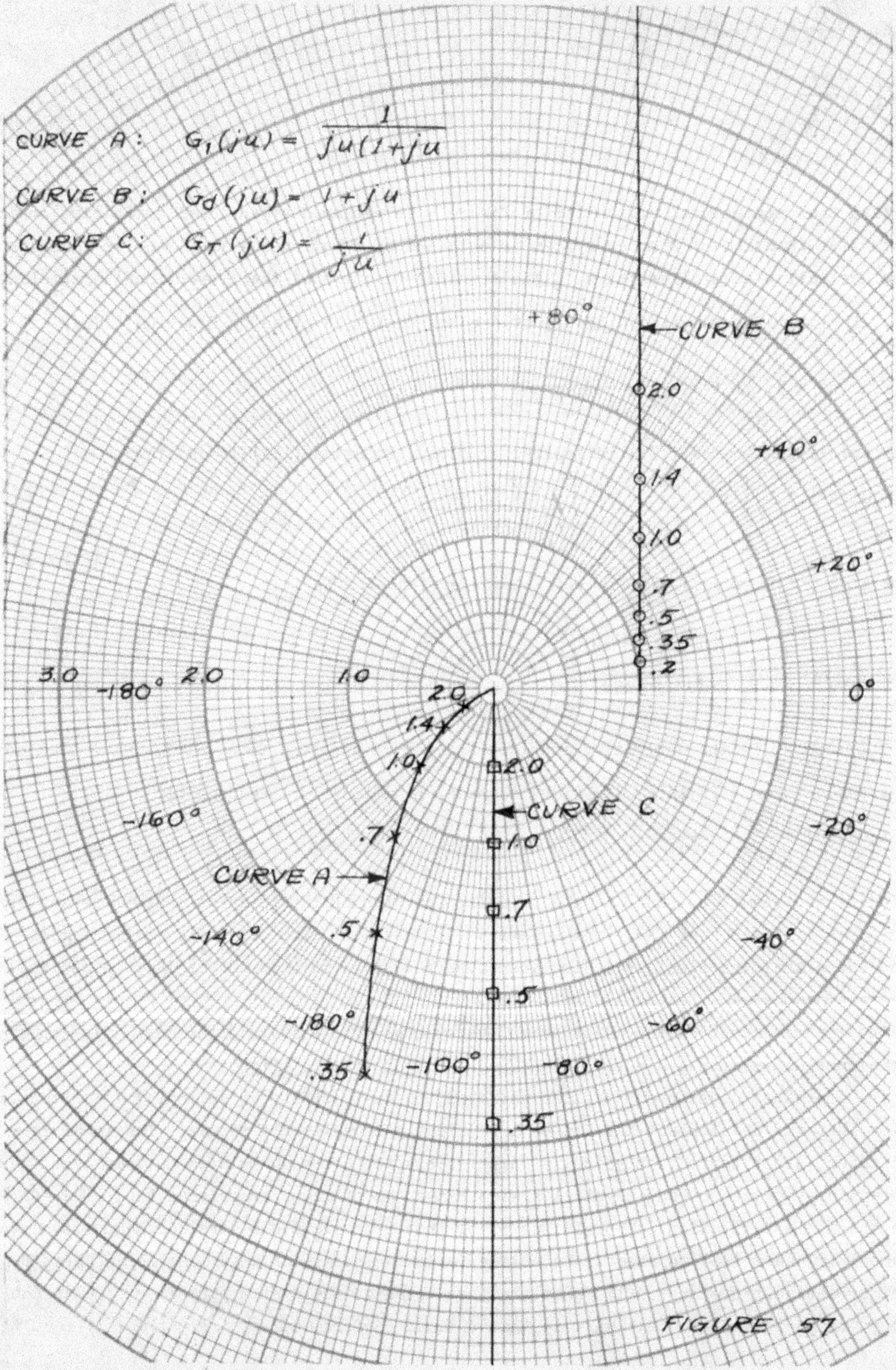

CURVE A: $G_1(ju) = \frac{1}{ju(1+ju)}$
CURVE B: $G_d(ju) = 1 + ju$
CURVE C: $G_T(ju) = \frac{1}{ju}$
+80°
CURVE B
2.0
+40°
1.4
1.0
+20°
.7
.5
.35
.2
3.0
-180°
2.0
1.0
0°
2.0
1.4
1.0
.7
CURVE A
.5
.35
2.0
CURVE C
1.0
.7
.5
.35
-160°
-20°
-140°
-40°
-180°
-60°
-100°
-80°

FIGURE 57

answered in order to obtain a physical design, and the second question must be answered if a decision is to be made as to whether or not that design will satisfactorily meet the application requirements. These questions cannot be answered by any derivation that leads to the conclusion that the speed of response of the resulting system can be increased without limit, because such a result cannot correspond to a physically realizable system.

The reason for the fallacy in the preceding discussion is that the ideal derivative controller assumed in the analysis cannot be obtained physically, since no physical device can be made to yield an output proportional to the mathematical derivative of the input. If correct conclusions are to be drawn, the assumption that the derivative controller is ideal and possesses a response specified by Equation (168) is not allowable, and it is necessary to express the defining equation of the controller in a manner that is in more harmony with physical realities. Accordingly, the assumption that such a controller is ideal will no longer be made, and only physical devices will be employed as dynamic-error compensators.

Basic Lead-Controller

A physical device that is commonly employed to simulate a derivative controller is that illustrated by figure 58, and termed in this paper a basic lead-controller.

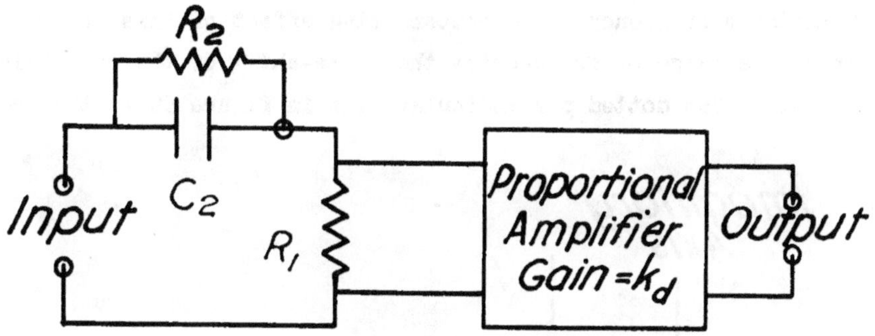

FIGURE 58

The transfer-function, $K_{da}G_{da}(s)$, of this controller is

$$K_{da}G_{da}(s) = \frac{k_d R_1}{R_1 + \dfrac{R_2}{1 + R_2 C_2 s}} \tag{188}$$

Equation (188) can be written

$$K_{da}G_{da}(s) = \frac{R_1 k_d}{R_1 + R_2} \frac{1 + R_2 C_2 s}{1 + \dfrac{R_1 R_2 C_2 s}{R_1 + R_2}} \tag{189}$$

Let $\dfrac{R_1 R_2 C_2}{R_1 + R_2} = \tau_d$, the network time constant. (190)

Let $\dfrac{R_1 + R_2}{R_1} = \alpha_d$, the network attenuation constant. (191)

Equation (189) becomes

$$K_{da}G_{da}(s) = \frac{k_d}{\alpha_d}\;\frac{1 + \tau_d\alpha_d s}{1 + \tau_d\; s}\;, \tag{192}$$

in which

$$K_{da} = \frac{k_d}{\alpha_d}\;, \tag{193}$$

$$G_{da}(s) = \frac{1 + \tau_d\alpha_d s}{1 + \tau_d s}\;. \tag{194}$$

Equation (192), as a function of the frequency, becomes,

$$K_{da}G_{da}(j\omega) = \frac{k_d}{\alpha_d}\;\frac{1 + \tau_d\alpha_d j\omega}{1 + \tau_d\; j\omega} \tag{195}$$

The transfer-locus of this controller is a semicircle constructed as shown in figure 59. At zero frequency, the transfer-vector has a relatively small magnitude and zero phase-shift; as the frequency increases, the vector magnitude increases, and the phase-angle of the vector increases in a positive direction to a maximum, beyond which the phase decreases until it again reaches zero at infinite frequency. The compensating effect of this controller is attributable to the fact that over a range of frequencies the phase-shift of the controller <u>increases positively</u> with frequency. The dotted perpendicular line in figure 59 is the locus of an ideal

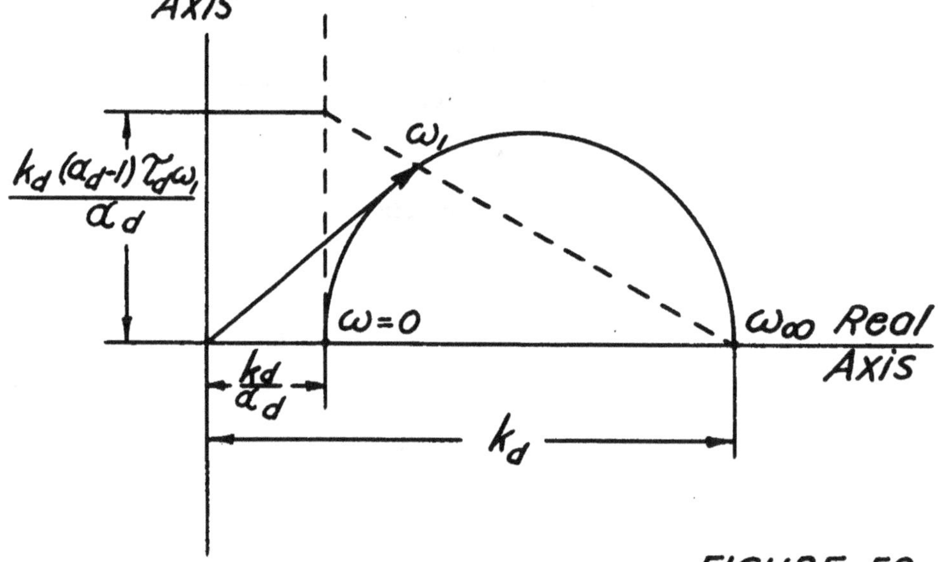

FIGURE 59

derivative controller, and it is seen that the locus of the physical controller approximates that of the ideal for those frequencies at which the semicircle and the perpendicular line approach tangency. The approximation of the ideal locus by the semicircle improves as the attenuation constant increases.

The ratio $\frac{k_d}{\alpha_d}$ is the value of the transfer-function of the lead-controller at zero frequency. It is the portion of the controller output that is proportional to the input and corresponds to the proportional term, K_d, of Equation (169). The factor $\frac{k_d}{\alpha_d}$ is also the gain factor of the lead-controller and is fixed by the requirements of the servomechanism. Since the ratio $\frac{k_d}{\alpha_d}$ is fixed and since the degree of approximation to an ideal derivative controller is proportional to the size of the attenuation constant, α_d, close approximation to the ideal is obtained only by utilizing a high value of k_d. Thus, a physical circuit that approximates a derivative controller with a high degree of accuracy is obtained only by employing a high gain amplifier.

The importance of the size of the attenuation constant is emphasized if the maximum positive phase-angle of the controller is calculated. As pointed out earlier, the compensating effect of the controller is due to the fact that the controller introduces a positive phase-shift into the system, thus cancelling a portion of the negative phase-shift introduced by the inertia and viscous damping present in a physical system. Therefore, the merit of the controller to a great extent is proportional to the maximum positive phase-angle that the controller can introduce into the system. The geometry of figure 59 makes it clear that this maximum phase-angle, which occurs at the frequency at which the transfer-vector is tangent to the semicircle, increases as α_d, the attenuation constant, increases. The maximum angle, ϕ_m, is given by Equation (196):

$$\phi_{dm} = \sin^{-1} \frac{\alpha_d - 1}{\alpha_d + 1} \qquad (196)$$

This maximum positive phase-shift occurs when

$$\tau_d \omega_p = \frac{1}{\sqrt{\alpha_d}} , \qquad (197)$$

in which ω_p is the required frequency. A table of maximum phase-shift, ϕ_{dm}, for various values of α_d the network attenuation constant together with other pertinent data is given in Table 3.

TABLE 3

α_d	ϕ_{dm}	$\tau_d \omega_p$	Ratio $\frac{G(j\omega_p)}{G(j0)}$
2.0	19°	.707	1.41
2.5	25°	.633	1.58
3.0	30°	.577	1.73
5.0	42°	.447	2.24
10.0	55°	.316	3.16
20.0	65°	.224	4.47
100.0	79°	.10	10.0
1000.0	86°	.032	31.6

The maximum angle, ϕ_{dm} , approaches 90° for infinite α_d but increases very slowly for values of α_d greater than ten.

Type I Servo with Basic Lead-Controller

Suppose the effect of combining the basic lead-controller, figure 58, with the Type I servo of figure 9 is investigated. The transfer-function of the combined system is obtained from Equation (176) by replacing k_p, the gain constant of the proportional amplifier of the Type I servo, by the transfer-function, Equation (192), of the basic lead circuit. The result is Equation (198).

$$K_T G_T(s) = \frac{k_d}{\alpha_d} \frac{1 + \tau_d \alpha_d s}{1 + \tau_d s} \frac{1}{f_L s(\tau_L s + 1)} \tag{198}$$

If the transformation

$$\tau_L s = s' \tag{177}$$

is made, Equation (198) becomes

$$K_T G_T(s') = \frac{k_d \tau_L}{\alpha_d f_L} \frac{1 + \frac{\tau_d \alpha_d}{\tau_L} s'}{1 + \frac{\tau_d}{\tau_L} s'} \frac{1}{s'(s' + 1)} \tag{199}$$

Let $$\frac{\alpha_d \tau_d}{\tau_L} = R_{dL} \tag{200}$$

Then $$K_T G_T(s') = \frac{k_d \tau_L}{\alpha_d f_L} \frac{1 + R_{dL} s'}{1 + \frac{R_{dL}}{\alpha_d} s'} \frac{1}{s' (s' + 1)} \tag{201}$$

The ratio R_{dL} may be adjusted at will by varying the time constnat, τ_d , of the derivative controller. If R_{dL} is made equal to unity, Equation (201) reduces to

$$K_T G_T(s') = \frac{k_d \tau_L}{\alpha_d^2 f_L} \frac{1}{\frac{s'}{\alpha_d} (1 + \frac{s'}{\alpha_d})} \tag{202}$$

In terms of the frequency, $u = \tau_L \omega$, Equation (202) is

$$K_T G_T(ju) = \frac{k_d \tau_L}{\alpha_d^2 f_L} \frac{1}{j \frac{u}{\alpha_d} (1 + j \frac{u}{\alpha_d})} \tag{203}$$

A comparison of Equation (203) and Equation (181), the expression for the transfer-function of the servo with a proportional instead of a derivative controller, reveals a very interesting fact: The frequency-dependent portions of the two expressions have the same form except that the frequency, u, of Equation (203) is modified by the factor, $\frac{1}{\alpha_d}$. Since the frequency-dependent portions of both systems are the same except for a scale change in the frequency, the shape of the transfer-loci of the two systems will be identical. Both loci will have the same position relative to the point (-1 + j0) and therefore, the proper numerical values of the gain factors are equal. However, the actual gain setting of the proportional amplifier is different in the two cases, and is a function of α_d for the servo with the lead-controller.

The nature of the servo output for the compensated system is the same, except that the time-scale of the latter system is shrunk by the factor α_d. That is, if the transient in the original system caused by a particular input required one second to decay to, say 10 per cent of its initial value, the transient in the lead-controller system excited by the same input would require only $\frac{1}{\alpha_d}$ second. If α_d were ten, there would be a ten-fold improvement in response.

At this point it might be concluded that in this example the servo response is augmented by the factor α_d by employing a lead-controller; and that it is only necessary to choose a controller in which the lead-network constant, α_d, is sufficiently large that the response of the servo is adequately fast to meet the specifications of whatever application is being considered. Such a conclusion is valid for a true Type I servo. Actually, however, no physical servo is as simple as a Type I servo, although in certain problems the actual characteristic of the physical system being studied may be approximated by the transfer-function of a Type I system. As long as the approximation is valid, the previous conclusion is correct. However, it will be shown that in general it is unsafe to approximate a physical servo system by a Type I representation if a lead-controller is to be employed with the system. In order to develop this matter more thoroughly, values of the ratio R_{dL} (see Equation (200)) other than unity will be investigated.

The effect upon the servo performance of varying the ratio R_{dL} is demonstrated by plotting the loci of the frequency-dependent portion of the transfer-function, Equation (201) for a range of values of R_{dL}. The function to be investigated is derived from Equation (201).

$$G_T(ju) = \frac{1 + jR_{dL}u}{1 + j\frac{R_{dL}u}{\alpha_d}} \frac{1}{(1 + ju)\,ju} \qquad (204)$$

The $G_T(ju)$ loci for a series of values of R_{dL} and a particular value of α_d are plotted in figure 60. As the magnitude of R_{dL} is decreased, the portion of the locus corresponding to low frequencies approaches more closely to the negative real axis, but that portion of the locus corresponding to higher frequencies is rotated counter-clockwise and does not approach the

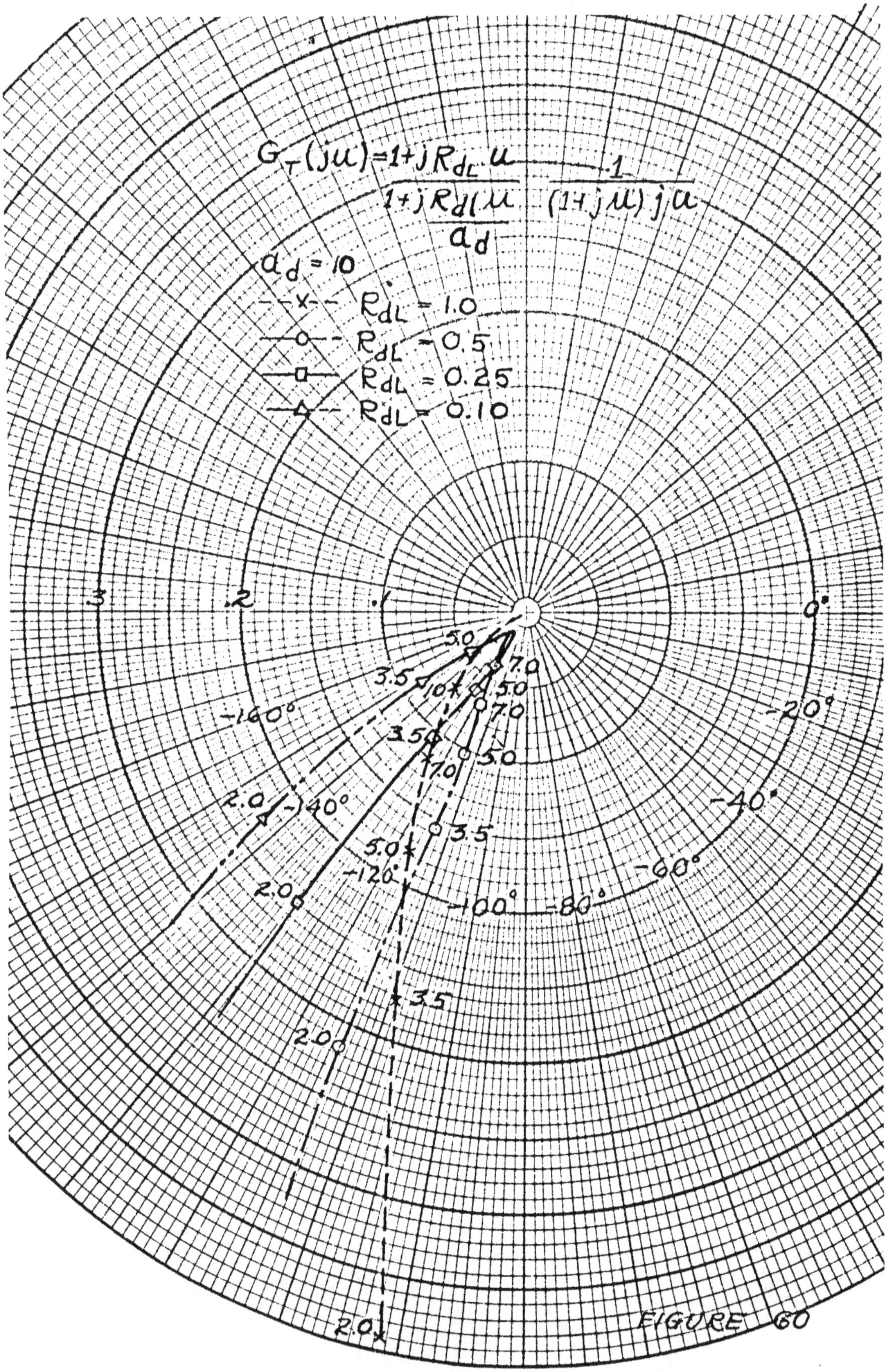

$G_T(j\mu) = \frac{1 + jR_{dL}\mu}{1 + j\frac{R_{dL}\mu}{a_d}} \cdot \frac{1}{(1+j\mu)j\mu}$
$a_d = 10$
--x-- $R_{dL} = 1.0$
--o-- $R_{dL} = 0.5$
--□-- $R_{dL} = 0.25$
--△-- $R_{dL} = 0.10$
.3
.2
.1
0°
-20°
-40°
-60°
-80°
-100°
-120°
-140°
-160°

FIGURE 60

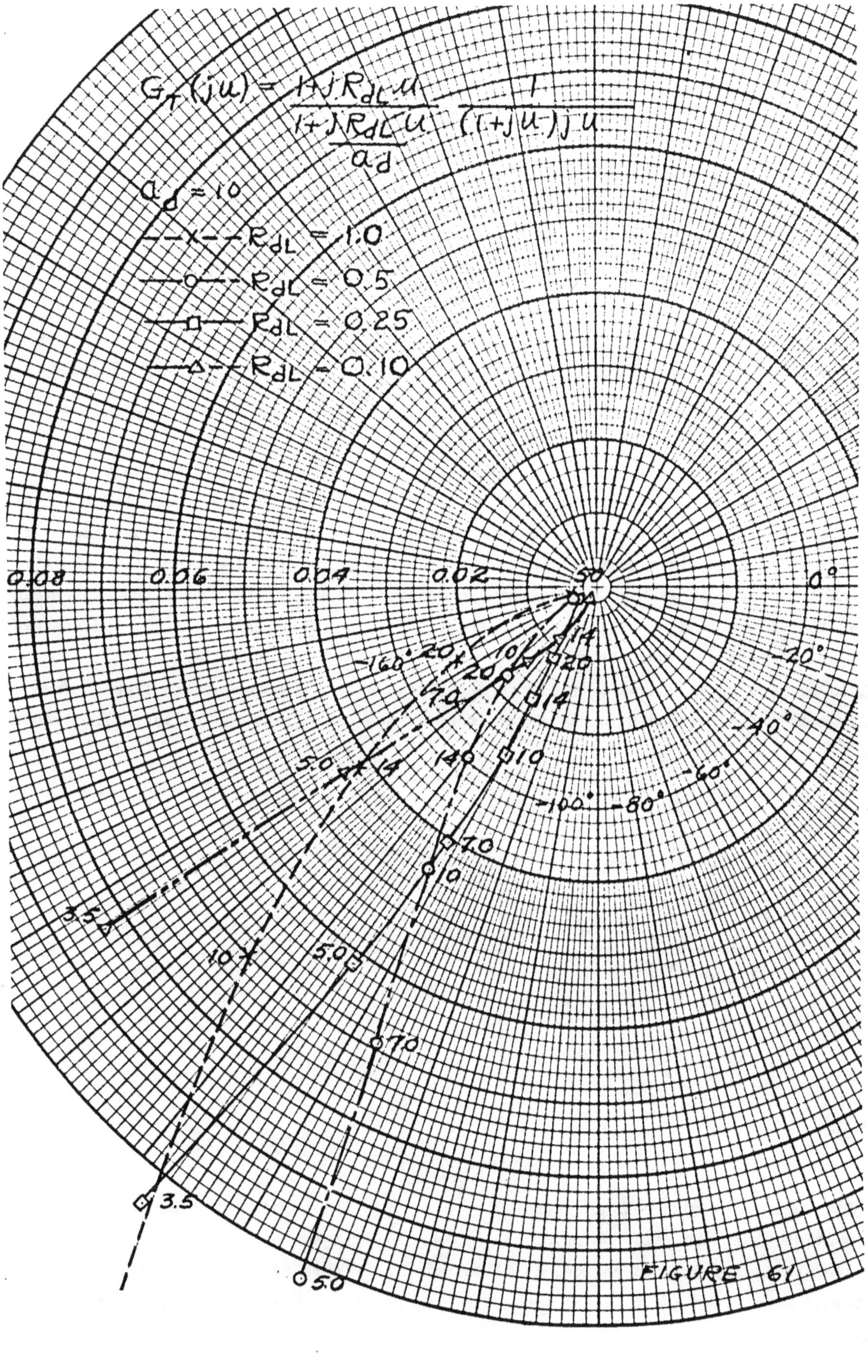
$G_T(ju) = \frac{1+jR_{dL}u}{1+\frac{jR_{dL}u}{a_d}} \frac{1}{(1+ju)ju}$
$a_d = 10$
--x-- $R_{dL} = 1.0$
—o— $R_{dL} = 0.5$
—□— $R_{dL} = 0.25$
—△— $R_{dL} = 0.10$
0.08
0.06
0.04
0.02
0°
-20°
-40°
-60°
-80°
-100°
-160°
FIGURE 61

real axis so closely. Therefore, if the gain factor, K_T, is increased as R_{dL} is decreased, the servo response remains well-damped. Moreover, the resonant frequency of the system increases as R_{dL} is decreased and the gain, K_T , correspondingly increased. Thus, it is seen that decreasing the magnitude of the ratio, R_{dL}, results in improved servo performance if accompanied by corresponding increases in gain. Figure 61 is a plot to a larger scale of the same G(ju) functions plotted in figure 60.

The logical question at this point is how small can the term, R_{dL}, be made and still result in improved servo performance? The answer is that R_{dL} can be decreased without limit, and that the servo performance will improve in proportion. This conclusion results from the fact that the transfer-locus of the original Type I servo (curve A, figure 57) approaches the origin along the negative real axis, and no matter how large the frequency, the phase of the transfer-function never becomes more negative than 180°. On the other hand, the location in the frequency spectrum of the band in which the lead-controller introduces positive phase-shift can be chosen indiscriminately by proper adjustment of the ratio R_{dL}. Therefore, the locus of the combined system (Type I servo with lead-controller) can be shifted counter-clockwise by an angle proportional to the maximum positive phase-shift of the lead-controller, and this positive shift can be made to occur at as high a frequency as desired. The gain factor of the system can be so adjusted that the positive shift occurs in the vicinity of the point (-1 + j0) and the system will be well-damped provided the positive phase-shift is sufficiently large.

This process is illustrated by figure 62. Curve A is the locus of the Type I servo at very large frequencies, and curve B is the locus of the basic lead-controller. The transfer-

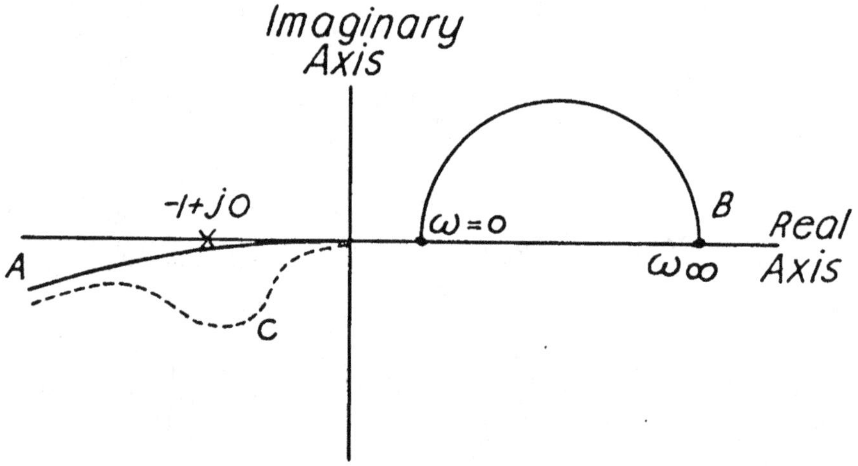

FIGURE 62

locus of the combined system is illustrated by curve C. The combined system is stable and well-damped provided that the gain factor of the system is so adjusted that the point (-1 + j0) lies within the positively shifted region, as illustrated in figure 62. The shapes of the loci are distorted to make clear the proceding reasoning but not to such an extent as to lead to erroneous conclusions.

The preceding reasoning is corroborated by the following more rigorous analysis. The transfer-function of the Type I servo with lead-controller is

$$K_T G_T(ju) = \frac{\tau_L k_d}{\alpha_d f_L} \frac{1 + jR_{dL} u}{1 + j \frac{R_{dL} u}{\alpha_d}} \frac{1}{(1 + ju)\, ju} \tag{205}$$

At very large frequencies (205) can be written:

$$K_T G_T(ju) = \frac{\tau_L k_d}{\alpha_d f_L} \frac{1}{(ju)^2} \frac{1 + jR_{dL}u}{1 + j \frac{R_{dL}u}{\alpha_d}} \tag{206}$$

In Equation (206), let

$$R_{dL}u = v \tag{207}$$

$$K_T G_T(jv) = \frac{\tau_L k_d R^2_{dL}}{\alpha_d f_L} \frac{1 + jv}{(1 + j \frac{v}{\alpha_d})\, (jv)^2} \tag{208}$$

The gain factor of Equation (208) is

$$K_T = \frac{\tau_L R^2_{dL} k_d}{f_L \alpha_d}, \tag{209}$$

and the frequency dependent portion, $G_T(jv)$ is

$$G_T(jv) = \frac{1 + jv}{(1 + j \frac{v}{\alpha_d})\, (jv)^2} \tag{210}$$

A series of loci of (210) for three values of the lead-network attenuation, α_d, are plotted to two scales in figures 63 and 64. The curves disclose that a value of α_d of ten or more is sufficiently large to ensure stable servo operation. The value of the gain factor, K_T, and the resonant frequency, v_m, can be determined in the usual manner. Thus, if α_d is ten, the following values of these constnats can be found:

$$K_T = 9.5 \tag{211}$$

$$v_m = 7.0$$

From relations (207) and (209),

$$u_m = \frac{v_m}{R_{dL}} = \frac{7.0}{R_{dL}} \tag{212}$$

$$k_d = \frac{K_T f_L \alpha_d}{\tau_L R^2_{dL}} = \frac{95\ f_L}{\tau_L R^2_{dL}}$$

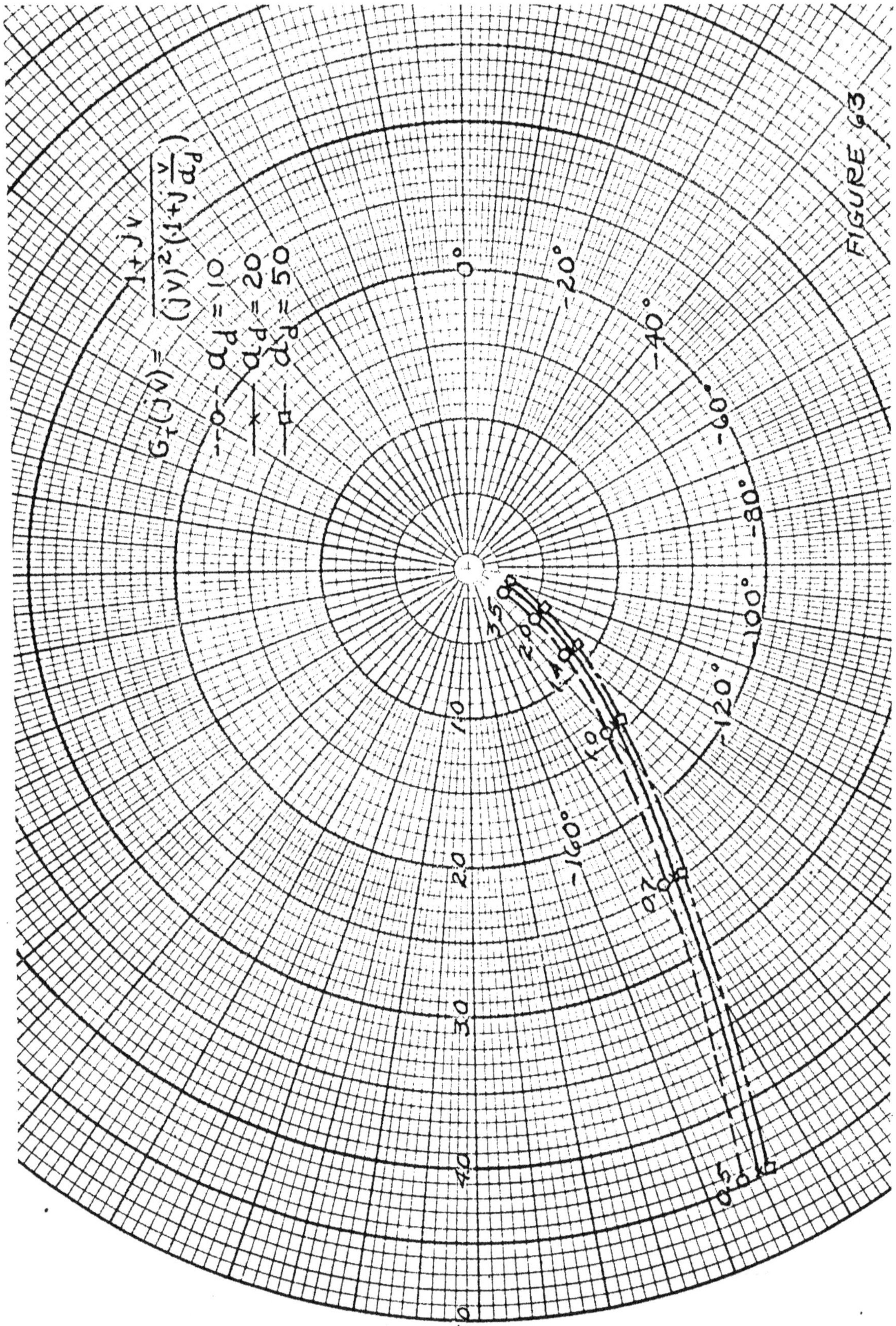

$G_\tau(jv) = \frac{1+jv}{(jv)^2(1+j\frac{v}{\alpha_d})}$
$\alpha_d = 10$
$\alpha_d = 20$
$\alpha_d = 50$
0°
-20°
-40°
-60°
-80°
-100°
-120°
-160°
1.0
2.0
3.0
4.0
5.0

FIGURE 6.3

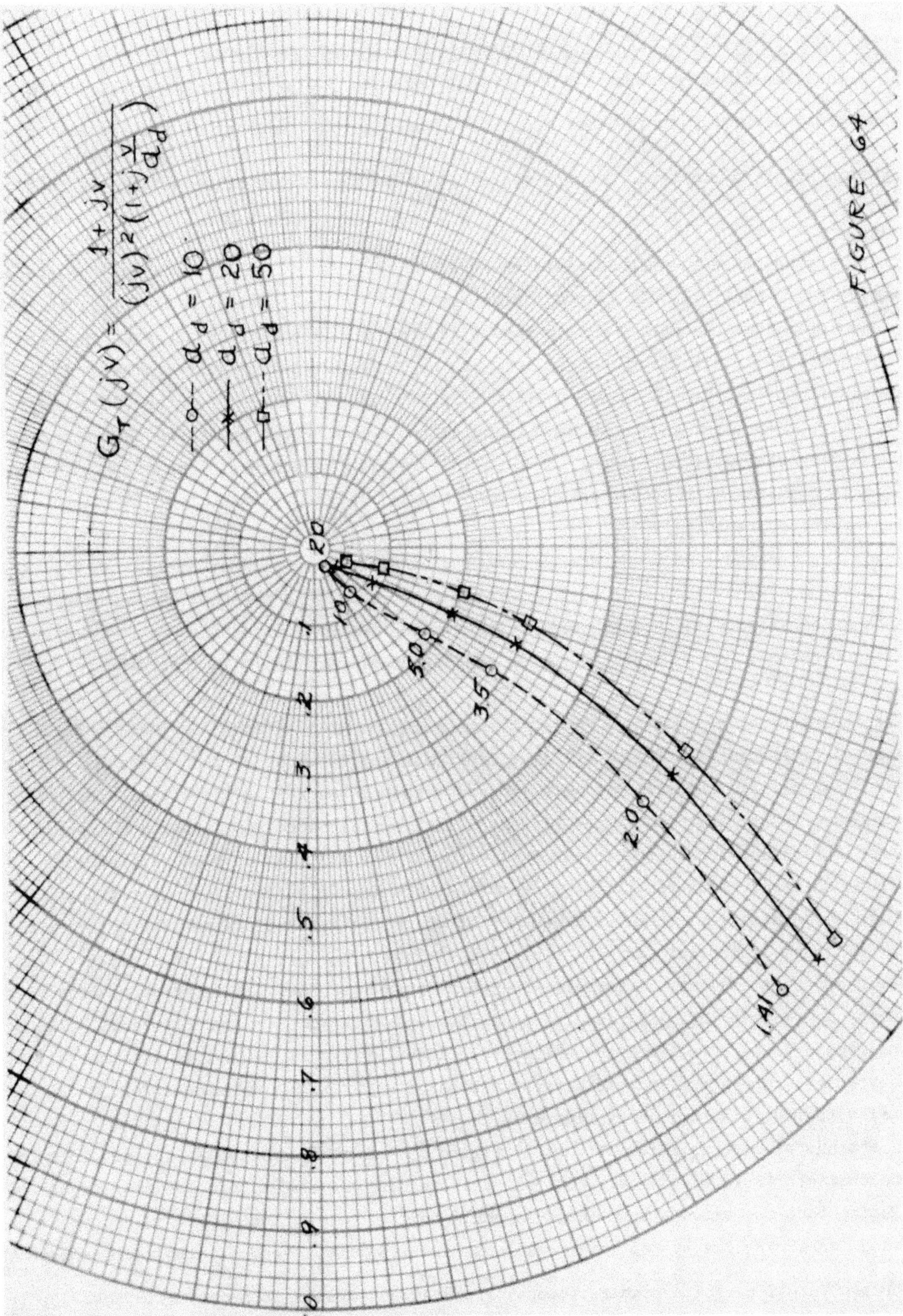
$G_r(jv) = \frac{1 + jv}{(jv)^2 (1 + j\frac{v}{\alpha_d})}$
$\alpha_d = 10$
$\alpha_d = 20$
$\alpha_d = 50$
20
10
5.0
3.5
2.0
1.41
0
.1
.2
.3
.4
.5
.6
.7
.8
.9

FIGURE 64

Limitations Upon the Transfer-Functions of Systems with Lead-Controllers.

From Equation (212) are obtained the conclusions that, provided the gain R_{dL} of the controller is varied inversely to the square of R_{dL} , the natural frequency of the system varies inversely with R_{dL}, and since R_{dL} can be varied at will, the system consequently can be made as fast as desired.

Thus, a conclusion is reached that the speed of response of the servo system can be made indefinitely fast. Such a result is obviously incorrect for any physical system, and it is important to uncover the reasons why such a result is obtained in the preceding example. There are two fundamental facts involved: the first is the fact that the negative phase-shift of the transfer-function of the Type I servo never exceeds 180° no matter how high the frequency; the second is the fact that the frequency band in which the slope of the phase-frequency characteristic is positive can be placed at any point in the frequency spectrum. That is, the frequency at which maximum phase-shift occurs can be chosen at will. The above two facts are the underlying reasons for the preceding result.

Actually, the phase of the transfer-function of every physical servomechanism will exceed 180° lag as the frequency increases without limit; furthermore, no physical lead-controller possesses such a characteristic that the frequency at which maximum phase-shift occurs can be chosen without restriction. These two points are discussed more fully in the following paragraphs.

The lead-controller of figure 58 has been assumed to have a transfer-function whose locus is illustrated by figure 59. This locus is correct for the elements of the network of figure 58. However, a physical resistance always possesses associated series inductance and shunt capacitance, and a physical capacitance possesses series resistance and inductance. These parasitic elements were not considered in the derivation of the semicircle locus of figure 59. If their effect had been considered, it would have been found sufficient to so alter the transfer-locus of the lead-controller that the frequency at which the maximum phase-shift occurs could no longer be chosen indiscriminately.

The basic fact is that the transfer-function of every physical network approaches zero as the frequency approaches infinity, and the parasitic elements that are associated with the lead-network of figure 58 will so alter the semi-circle locus of figure 59 that it conforms with this basic requirement. Therefore, the transfer-function of a physical lead-controller possesses a locus similar to that illustrated by figure 65.

If a transfer-locus similar to that of figure 65 had been employed in the preceding example as the characteristic of the lead-controller instead of the semi-circle locus of figure 59, the conclusion that the servo system response could be improved without limit by decreasing the term R_{dL} and increasing the system gain would not have resulted. From this it would appear that an exact representation of the transfer-locus of a lead-controller is necessary. However, if the lead-controller is an electrical network, circuit elements can be employed that are sufficiently free of parasitic elements that the semicircle is an accurate representation of the transfer-locus over a frequency band extending into the radio frequencies. An accuracy over such a frequency range is far superior to the accuracy with which a Type I transfer-function represents a physical servomechanism over the same frequency range. As a matter of fact, the semicircle representation of the lead-controller is more accurate than almost any ordinary representation of a physical servomechanism.

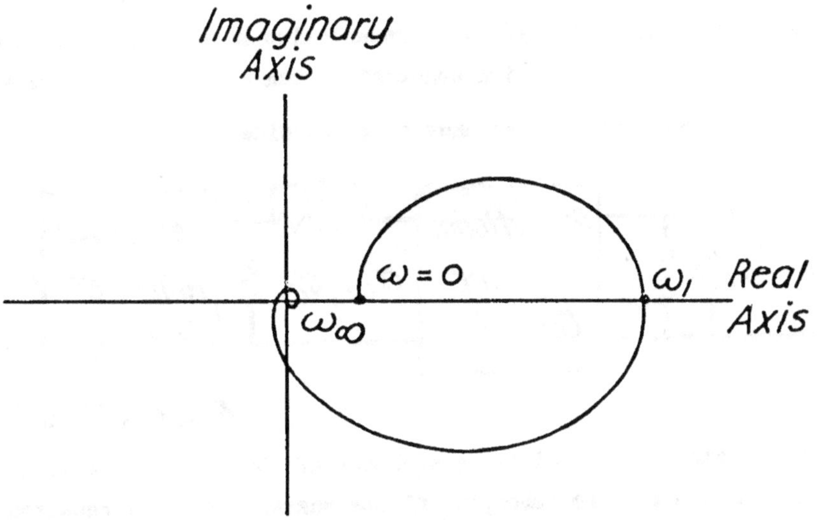

FIGURE 65

This brings up the second point for discussion: namely, that while the phase of the Type I function never exceeds -180°, the phase of the transfer-function of an actual servomechanism will always exceed that value as the frequency increases indefinitely. Thus, no physical servo system is so simple that it can be perfectly represented by a Type I transfer-function. For certain purposes, a physical system may be approximately characterized by this simple transfer-function, but such an approximation is invalid if the effect of employing a lead-controller with a large attenuation constant in the system is to be determined.

When a physical system is characterized by a mathematical relationship employing linear differential equations, (or their equivalent), that relationship is always an approximate one. The form and complexity of the relationship that is employed depends upon the accuracy with which it is desired to represent the response of the physical system. If the resonant frequencies of the system and the magnitude of the system response at those frequencies are predicted with a tolerable accuracy by the system representation, that representation is generally considered to be adequate.

It is always desirable to employ the simplest mathematical characterization of a system that will yield results of satisfactory accuracy. If certain types of servomechanisms are to employ proportional controllers only, a Type I representation may predict the resonant frequency and the system response at that frequency with quite satisfactory accuracy. If the resonant frequency of the physical system is ω_m, the Type I transfer-function must approximate the system transfer-function over a frequency range extending from zero to approximately ω_m.

The analysis on page 98 - 99 has shown that if a lead-controller with an attenuation constant of, say, two is employed with a Type I system, the resonant frequency of the compensated system is twice that of the original system. Therefore, if a Type I representation is to be used to characterize a system with a lead-controller whose attenuation constant is equal to two, the representation must be accurate up to a frequency equal to $2\omega_m$. Similarly, if the attenuation constant is equal to three, the Type I representation must be good up to a frequency equal to about $3\omega_m$. However, if the attenuation constant is ten or larger, the Type I representation must be approximately correct over a frequency range as wide as that over which the semicircle

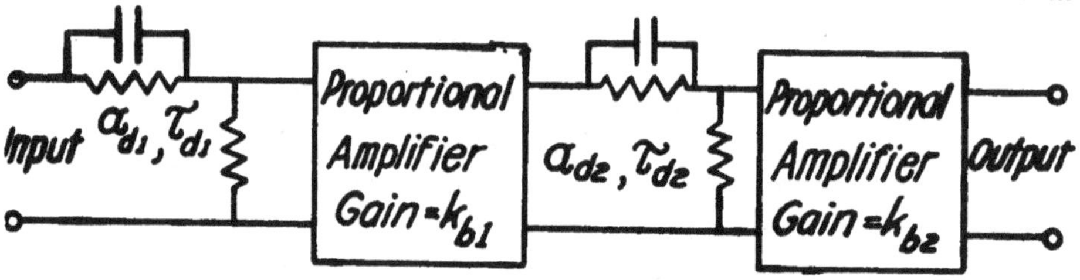

FIGURE 66

transfer-locus approximates the actual transfer-locus of the lead-controller. Since no physical servomechanism with a structure embodying finite masses can be represented as a Type I system over such a wide frequency range, it must be concluded that a more complex representation must be employed if a lead-controller with a large attenuation-constant is to be used.

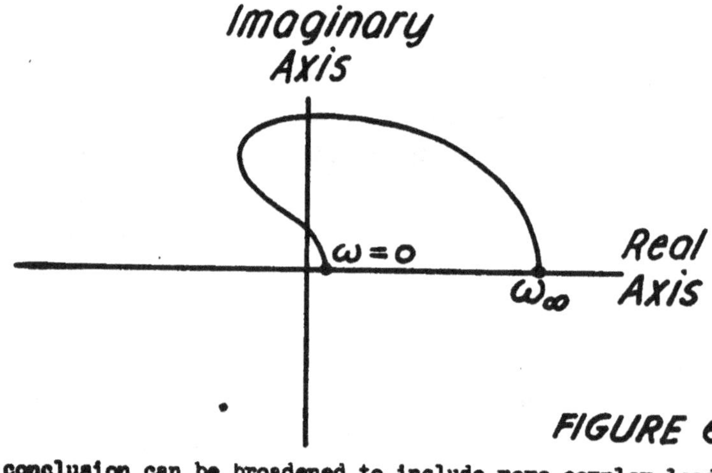

FIGURE 67

The preceding conclusion can be broadened to include more complex lead-controllers. For example, suppose a controller composed of two cascaded lead networks and isolating amplifiers such as illustrated by figure 66 is to be studied. The transfer-locus of such a controller will have twice the phase-shift of a single section lead-controller if the two sections are identical, and might have a transfer-locus such as figure 67. If the maximum positive phase-shift of such a controller is larger than about 135°, and the effect of embodying such a device in a system characterized by a third-order transfer-function is studied, the same conclusion would again be drawn that such a servo could be made as fast as desired. The reason for this conclusion is similar to that of the previous example: namely, that the maximum negative phase-shift of the third-order transfer-function is 270°; if the maximum positive phase-shift of the controller is about 135° or more, the transfer-locus of the complete system can be shifted across the negative real axis and the system gain can be so adjusted that the system is stable and well damped; and, finally, by reducing the time constant of the cascaded controller, the frequency at which the phase-shift occurs can be set at any desired value Figure 68 is an illustration of the transfer-loci of the components of such a system. Curve A

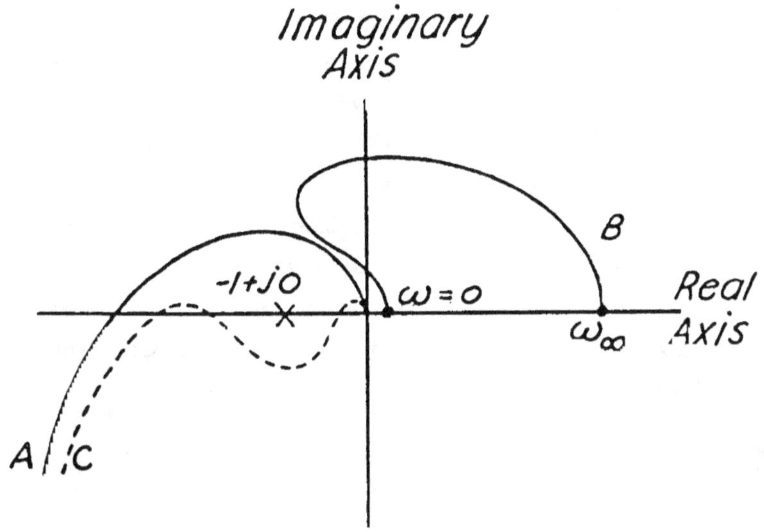

FIGURE 68

is a third-order transfer-function locus; curve B is the transfer-locus of the servo lead-controller; and curve C is the locus of the complete system. The dip in the locus, curve C, can be adjusted to occur at any desired frequency. The scale is distorted somewhat in order to clarify the situation.

From the reasoning of the preceding paragraphs, it is obvious that the system with which a lead-controller of transfer-locus similar to figure 67 is to be used must be characterized by at least a fourth-order transfer-function if significant design results are to be obtained.

The reasoning behind such a conclusion is exactly the same as that previously formulated for the basic lead-controller and the Type I transfer-function. Briefly, if the maximum phase-shift of the lead-controller is greater than about 135°, the third-order transfer-function must be an accurate representation of the physical servomechanism over a frequency range as wide as that over which the semi-circle transfer-locus approximates the actual locus of one of the stages of the double-stage lead-controller. Since this is obviously impossible, it is necessary to employ at least a fourth-order transfer-function to represent a servomechanism with which such a lead-controller is to be employed.

If the maximum phase-shift of the double-stage lead-controller is about 100°, the third-order representation may possibly be used to characterize the physical system. The test is to determine whether or not the third-order transfer-function is an accurate representation of the physical system over a frequency range extending to approximately the resonant frequency of the third-order system with compensating lead-controller.

It is now possible to extend the above reasoning to the analysis of all systems comprising lead-controllers. In every case the order of transfer-function employed to characterize the physical system must be such that its negative phase-shift at infinite frequency is numerically greater by at least 160° to 180° than the maximum positive phase-shift of the lead-controller.

Third-Order Transfer-Function Servomechanism with Basic Lead-Controller

The correct procedure to be followed in the design of a basic lead-controller to be embodied in a physical servo is clarified if an example is carried through. The preceding sections of this chapter have shown that certain precautions must be taken in the choice of the transfer-function that is to characterize the servo with which the lead-controller is to be used. In particular it has been shown that a third-order transfer-function is the lowest order function that will correctly characterize a servo with which a basic lead-controller with a large attenuation constant is to be used. Assume, therefore, that the third-order function considered on page 54 characterizes a physical system with a proportional controller, and a basic lead-controller whose transfer-locus is given by figure 59 is to be employed with that system. The third-order transfer-function is given by Equation (115) the symbols of which are defined by the text beginning with page 55.

$$K_3G_3(s) = \frac{k_pk_m}{J_d}\,\frac{1}{s(s^2 + 2\zeta_3\omega_0 s + \omega_0^2)} \tag{115}$$

The transfer-function of the lead-controller is given by Equation (192), the symbols of which are defined in the text beginning with page 91.

$$K_{da}G_{da}(s) = \frac{k_d}{\alpha_d}\,\frac{1 + \tau_d\alpha_d s}{1 + \tau_d s} \tag{192}$$

The transfer-function of the combined system is obtained by replacing the transfer-function, k_p, of the proportional controller of the third-order system by the transfer-function, Equation (192), of the lead-controller. The transfer-function thus obtained is denoted by $K_TG_T(s)$.

$$K_TG_T(s) = \frac{k_dk_m}{\alpha_dJ_d}\,\frac{1}{s(s^2 + 2\zeta_3\omega_0 s + \omega_0^2)}\,\frac{1 + \tau_d\alpha_d s}{1 + \tau_d s} \tag{213}$$

If the following substitutions are made,

$$s' = \frac{s}{\omega_0} \tag{116}$$

$$R_{d3} = \tau_d\omega_0 \tag{214}$$

Equation (123) becomes

$$K_TG_T(s') = \frac{k_dk_m}{\alpha_dJ_d\omega_0^3}\,\frac{1}{s'(s'^2 + 2\zeta_3 s' + 1)}\,\frac{1 + \alpha_dR_{d3}s'}{1 + R_{d3}s'} \tag{215}$$

The transfer-function of Equation (126) can be separated into one portion that is invariant with frequency and a second portion that is frequency dependent:

$$K_T = \frac{k_d k_m}{\alpha_d J_d \omega_0^3} \tag{217}$$

$$G_T(ju) = \frac{1}{ju(-u^2 + 2j\zeta_3 u + 1)} \quad \frac{1 + j\alpha_d R_{d3} u}{1 + jR_{d3} u} \tag{218}$$

The problem is so to choose the gain factor, K_T, and the adjustable parameters of G(ju) that optimum servo performance results.

The ratio, R_{d3} , is always adjustable; the controller attenuation constant, α_d, is generally adjustable, and the damping ratio, ζ_3 , may or may not be adjustable. For the purposes of this example the assumption is made that the damping ratio, ζ_3, is fixed and equal to unity, and the problem defined to be that of determining the improvement in servo response that may be achieved with a given value of controller attenuation, α_d, and a proper choice of the ratio, R_{d3}.

The approach to the problem of proper determination of the adjustable servo parameters is that employed in examples previously considered: namely, the loci of the G(ju) function is plotted for various values of the adjustable servo parameter and that value selected that results in an optimum locus from the point of view of the various factors listed on page 46. First, it is worth while to summarize the pertinent facts concerning the performance of the third-order system with proportional controller. The locus of the frequency-dependent portion of the transfer-function is plotted in figure 30, and the pertinent results are summarized in Table 1 for the case in which the maximum value of $\left|\frac{\theta_{o(j\omega)}}{\theta_i}\right|$ is limited to one and one-third. It is shown there that the proper gain factor of the system is 0.58 and the resonant frequency of the system is 0.60. These figures can be used as a reference to gauge the improvement secured when a lead-controller is employed with the system.

An approximate idea of the improvement in performance made possible by a lead-controller is gained by the following reasoning. The action of the lead-controller is to shift the G(ju) locus counterclockwise and thus increase the resonant frequency, u_m, for which $\left|\frac{\theta_{o(ju)}}{\theta_i}\right|$ is maximum. A large resonant frequency is desirable and therefore, it is desirable to secure as large an angular shift in the locus as possible. The discussion of the preceding chapter has shown that the maximum positive shift that the basic lead-controller can introduce is a function of the network attenuation constant, α_d , and table 3, page 97, presents a tabulation of the maximum phase-shift of this controller for various values of α_d. If the <u>shape</u> of the original servo locus were not altered by the action of the lead-controller, the action of the lead-controller would be equivalent to a frequency transformation in the original transfer-function. In that case, the frequency u_{mT} of the combined system could be found from the frequency u_{m3} of the original system, by determining the phase-angle of the lead-controller, and reading the frequency, u_{mT}, corresponding to this value of phase-shift on the locus of $G_3(ju)$. In equation form, this procedure is

$$\text{arc}\left[G_3(ju_{mT})\right] = \text{arc}\left[G_3(ju_{m3})\right] + \phi_{dm} \tag{219}$$

in which ϕ_{dm} is the maximum phase-shift of the lead-controller for the particular value of attenuation constant employed. For example, suppose that the attenuation of the lead-controller network is ten. By table 3, the maximum positive phase-shift of this controller is 55 degrees. The frequency, u_m, of the third-order function being considered (ζ_3 equals unity) is 0.6, and the phase-angle of $G_3(j0.6)$ is about 145 degrees. On the same locus, the frequency at which the phase reaches 200 degrees ($55^\circ + 145^\circ$) is about 1.4. Therefore, if the introduction of the lead-controller left the shape of the G(ju) function unchanged, the lead-controller would result in an increase of the resonant frequency, u_m, from 0.6 to 1.4. The natural frequency of the system would increase in about the same proportion.

Actually, the shape of the G(ju) locus is altered by the introduction of the lead-controller and the above procedure is not quite correct. However, under most circumstances the method yields an approximately correct answer and has the advantage of simplicity.

The approximation that the shape of the G(ju) locus is unchanged by the introduction of the lead-controller is based upon (1) the fact that for low frequencies the phase of the transfer-function of the lead-controller varies much more rapidly than the magnitude of the transfer-function, and (2) the fact that the phase of the transfer-function is more important in determining servo response than the magnitude. Actually, the shape of the transfer-function of a servo is unaltered by the introcution of a controller provided that the magnitude of the transfer-function of the controller is invariant with frequency. The action of such a controller is only to shift the frequency scale of the locus of the combined system from that of the original servo. If the variation in the magnitude of the transfer-function of the lead-controller is not large over the frequency range from zero to that at which maximum phase-shift occurs, the effect of the controller upon the shape of the locus of the original system will not be large, and the preceding calculation will be approximately correct.

Information concerning the relative change in the magnitude of the transfer-vector of the lead-controller as the frequency varies from zero to that value at which maximum positive phase-shift occurs is tabulated on page 97. The ratio of the magnitude of the transfer vector at the frequency, ω_p, at which maximum phase-shift occurs, to the magnitude at zero frequency is found to be

$$\frac{G(j\omega_p)}{G(j0)} = \sqrt{\alpha_d} \tag{220}$$

For values of α_d below approximately ten, the variation in the magnitude of the transfer-vector has a second-order effect.

The preceding analysis has made it clear that if the shape of the locus is unchanged by by the lead-controller, optimum controller adjustment is obtained by securing a maximum shift in the resonant frequency of the system. It has been shown that maximum shift is secured if Equation (219) holds. This relation will hold provided that the maximum phase-shift of the lead-controller occurs at the frequency, u_{mT}. The frequency at which maximum phase-shift occurs is given by Equation (197) and tabulated on page 97.

In the example being considered it has been found that $u_{mT} = 1.4$ if $\alpha_d = 10$. Thus

$$\tau_d \omega_{mT} = .316,$$

$$\tau_d \omega_0 \frac{(\omega_{mT})}{\omega_0} = .316,$$

$$R_{d3}\, u_{mT} = .316.$$

But $u_{mT} = 1.4$. Therefore

$$R_{d3} = \frac{.316}{1.4} \cong .23.$$

It is emphasized that the preceding calculations are approximations which are useful because of their simplicity and the fact that they enable an estimate of the system response to be made. In general they should be supplemented by plots of the transfer-loci of the system.

A family of loci for α_d equal to ten and various values of R_{d3} are plotted in figure 69. It is evident from these curves that a value of R_{d3} approximately equal to 0.2 yields the optimum locus. The value of 0.2 is very close to that predicted by the preceding analysis. The fact that the shape of the locus is altered by the lead-controller is clear from the curves of figure 69, but the change is not such to invalidate the approximations made heretofore. The pertinent data of the system are compiled in Table 4. It should be noted that the resonant frequency of the system is 1.3, instead of 1.4, as was predicted in the preceding estimate.

$$\left| \frac{\theta_o}{\theta_i}(ju) \right|_{max} = 1.33$$

R_{d3}	K_T	u_m
0	0.58	0.6
0.05	0.76	0.72
0.1	1.04	0.9
0.2	1.04	1.3
0.5	.46	1.3

TABLE 4

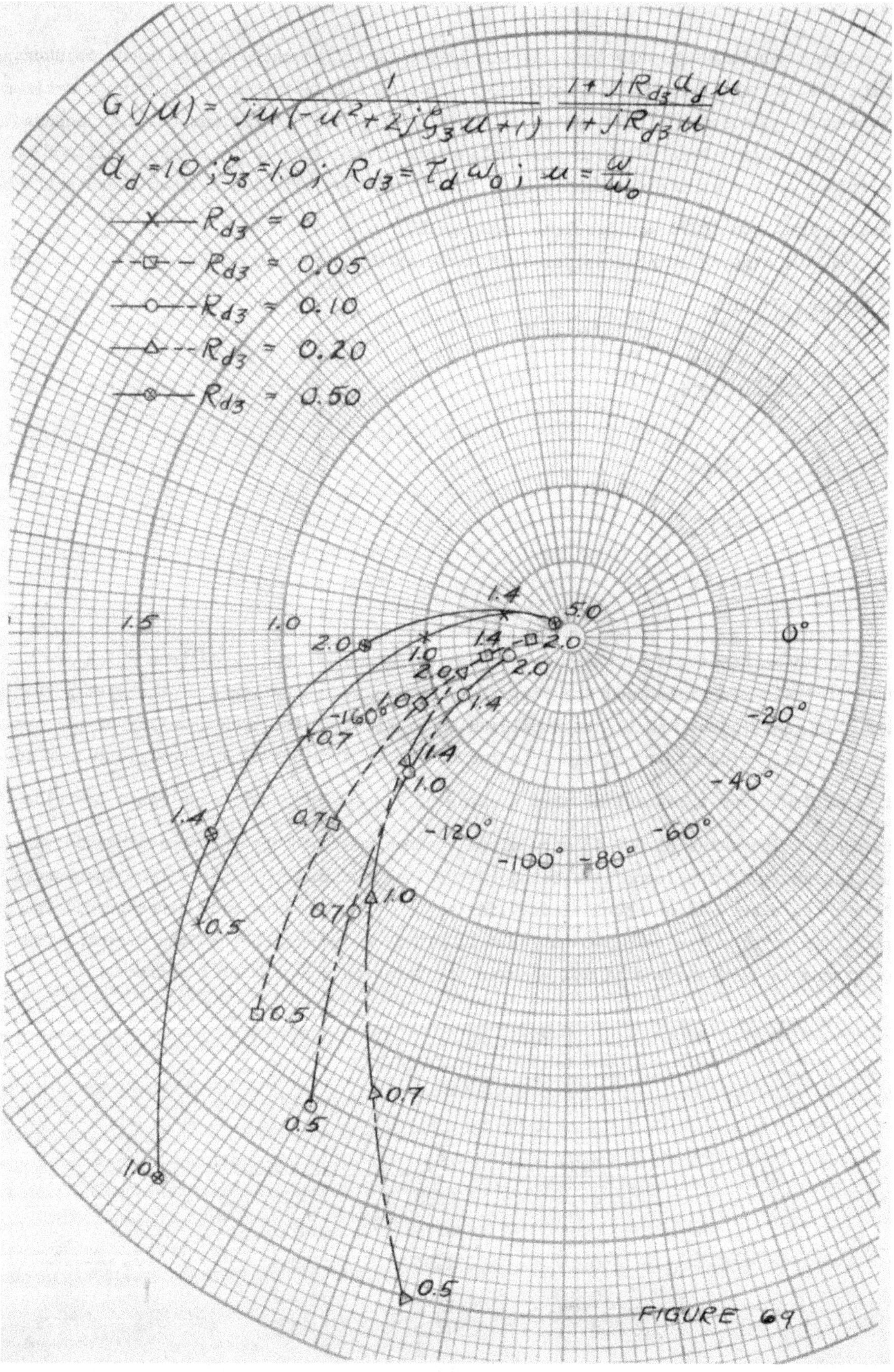

$G(j\mu) = \frac{1}{j\mu(-\mu^2+2j\zeta_3\mu+1)} \frac{1+jR_{d3}\alpha_d\mu}{1+jR_{d3}\mu}$
$\alpha_d = 10;\ \zeta_3 = 1.0;\ R_{d3} = \tau_d\omega_0;\ \mu = \frac{\omega}{\omega_0}$
$R_{d3} = 0$
$R_{d3} = 0.05$
$R_{d3} = 0.10$
$R_{d3} = 0.20$
$R_{d3} = 0.50$
1.5
1.0
0°
-20°
-40°
-60°
-80°
-100°
-120°
-160°

FIGURE 69

<u>Third-Order</u> <u>Transfer-Function</u> <u>Servo</u> <u>with</u> <u>Compound</u> <u>Lead-Controller</u>; Case 1: $\zeta_3 = 1.0$

In an early part of this chapter the suggestion was made that in some instances it is preferable to employ a double-stage lead-controller in place of the basic lead-controller discussed hitherto. One of the reasons that this is true is because the double-stage lead-controller is capable of introducing a larger value of positive phase-shift for the same total controller attenuation. A double-stage lead-controller is illustrated by figure 66. If the two networks in the circuit are identical, the maximum positive phase-shift of the complete controller is twice that of a single stage.

Table 5 compares the maximum phase-shift of the double-stage lead-controller with that of the basic lead-controller for equal values of overall attenuation constant.

Single-Stage Lead Controller		Double-Stage Lead Controller	
α_d	ϕ_{dm}	$\alpha_{d1} = \alpha_{d2} = \sqrt{\alpha_d}$	ϕ_{dm}
2.0	19°	$\sqrt{2}$	19.8°
3.0	30°	$\sqrt{3}$	31.1°
5.0	42°	$\sqrt{5}$	45°
10.0	55°	$\sqrt{10}$	62.6°
20.0	65°	$\sqrt{20}$	78.6°
100.0	79°	$\sqrt{100}$	110°
1000.0	86°	$\sqrt{1000}$	140°

TABLE 5

Design principles underlying the application of double-stage lead-controllers (and multi-stage lead-controllers) are effectively introduced by first considering the application of such a controller to a servo system to which a single-stage controller has been applied. Contrasts in the results emphasizes important design principles.

It has previously been shown that the maximum phase-shift, ϕ_{dm} , of a basic lead-controller is a function of the attenuation constant of the network. When two basic lead-controllers are cascaded to form a double-stage lead-controller, (see figure 66) the attenuation constant of the complete controller is the product of the attenuation constants of each individual section, and the phase-shift is the sum of the phase-shifts of each section. The double-stage lead-controller, therefore, has maximum phase-shift when the two sections are so adjusted that the maximum phase-shift of each section occurs at the same frequency. The difference between the maximum phase-shifts of the two types of controllers is especially pronounced for large values of attenuation constant. This is true because the phase-shift of the single-stage lead-controller cannot exceed 90 degrees, while that of the double-stage lead-controller can reach 180 degrees.

When a double-stage lead-controller is to be employed with a system represented by a third-order transfer-function, precautions must be taken in order that erroneous conclusions are not reached, since for large values of attenuation constant the maximum positive phase-shift of the double-stage lead-controller can approach 180 degrees while the negative phase-shift of the third-order transfer-function servo cannot exceed 270 degrees. The difficulties that may be encountered in such a situation have been discussed in the preceding section. However, in the following example only values of attenuation constant are considered for which the maximum positive phase-shift of the controller is less than 90 degrees.

The third-order transfer-function is given by Equation (115)

$$K_3G_3(s) = \frac{k_p k_m}{J_d} \frac{1}{s(s^2 + 2\zeta_3\omega_0 s + \omega_0^2)} \tag{115}$$

The transfer-function, $K_{db}G_{db}(s)$, of a double-stage lead-controller (see figure 66) in which the two sections are characterized by attenuation constants α_{d1} and α_{d2} and time constants τ_{d1} and τ_{d2} is

$$K_{db}G_{db}(s) = \frac{k_{b1}k_{b2}}{\alpha_{d1}\alpha_{d2}} \frac{1 + \tau_{d1}\alpha_{d1}s}{1 + \tau_{d1}s} \frac{1 + \tau_{d2}\alpha_{d2}s}{1 + \tau_{d2}s} \tag{221}$$

The transfer-function, $K_TG_T(s)$, of the combined system is found by replacing k_p, the term for the gain factor of the proportional amplifier in Equation (115), by the expression for $K_{db}G_{db}(s)$ Thus

$$K_TG_T(s) = \frac{k_{b1}k_{b2}k_m}{J_d\alpha_{d1}\alpha_{d2}} \frac{1}{s(s^2 + 2\zeta_3\omega_0 s + \omega_0^2)} \frac{1 + \tau_{d1}\alpha_{d1}s}{1 + \tau_{d1}s} \frac{1 + \tau_{d2}\alpha_{d2}s}{1 + \tau_{d2}s} \tag{222}$$

If the following substitutions are made,

$$s' = \frac{s}{\omega_0} \tag{116}$$

$$R'_{d3} = \tau_{d1}\omega_0 \tag{223}$$

$$R''_{d3} = \tau_{d2}\omega_0 \tag{224}$$

Equation (222) becomes

$$K_TG_T(s') = \frac{k_{b1}k_{b2}k_m}{J_d\alpha_{d1}\alpha_{d2}\omega_0^3} \frac{1}{s'(s'^2 + 2\zeta_3 s' + 1)} \frac{1 + R'_{d3}\alpha_{d1}s'}{1 + R'_{d3}s'} \frac{1 + R''_{d3}\alpha_{d2}s'}{1 + R''_{d3}s'} \tag{225}$$

The preceding example concerned itself with a particular value of damping ratio, ζ_3, namely, unity. In order to permit comparisons to be made, that some value is assigned to the damping ratio in Equation (225). In addition, the attenuation constants, α_{d1} and α_{d2}, are set equal to one another and their common value denoted by α_0 . The reason for this is made apparent by the form of Equation (230). Equation (225) becomes

$$K_TG_T(s') = \frac{k_{b1}k_{b2}k_m}{J_d\alpha_0^2\omega_0^3} \left[\frac{1}{s'(s'^2 + 2s' + 1}\right] \frac{1 + R'_{d3}\alpha_0 s'}{1 + R'_{d3}s'} \frac{1 + R''_{d3}\alpha_0 s'}{1 + R''_{d3}s'} \tag{226}$$

The portion of Equation (226) enclosed in the brackets is the frequency-dependent portion of the third-order transfer-function. This portion can be factored. Thus,

$$K_T G_T(s') = \frac{k_{b1} k_{b2} k_m}{J_d \alpha_0^2 \omega_0^3} \left[\frac{1}{s'(s'+1)(s'+1)} \right] \cdot \frac{1 + R'_{d3}\alpha_0 s'}{1 + R'_{d3}\, s'} \; \frac{1 + R''_{d3}\, \alpha_0 s'}{1 + R''_{d3}\, s'} \tag{227}$$

Let $R'_{d3}\,\alpha_0 = R''_{d3}\,\alpha_0 = 1$ (228)

Equation (227) reduces to,

$$K_T G_T(s') = \frac{k_{b1} k_{b2} k_m}{J_d \alpha_0^2 \omega_0^3} \; \frac{1}{s' \left(1 + \frac{s'}{\alpha_0}\right)\left(1 + \frac{s'}{\alpha_0}\right)} \tag{229}$$

Equation (229) can be written,

$$K_T G_T(s') = \frac{k_{b1} k_{b2} k_m}{J_d \alpha_0^3 \omega_0^3} \left[\frac{1}{\frac{s'}{\alpha_0} \left(1 + \frac{s'}{\alpha_0}\right) \left(1 + \frac{s'}{\alpha_0}\right)} \right] \tag{230}$$

The form of the bracketed portion of Equation (230) is precisely that of the bracketed portion of Equation (227), except the complex variable s' of Equation (230) is divided by α_0. Since the bracketed portion of Equation (227) is the G(s') function of the third-order system, the conclusion is reached that the form of the frequency response of the system of Equation (230) is precisely that of the third-order system except it is extended in the frequency scale by a factor of α_0. Thus, this development has culminated in a servo system with the same damping ratio but with α_0 times the natural frequency of the initial system. This type of compensation preserves the form of the transfer-locus of the original system and only changes the frequency scale.

Summarized, the pertinent facts of the new system are

$$K_T = \frac{k_{b1} k_{b2} k_m}{J_d \alpha_0^3 \omega_0^3} = 0.58 \qquad u_m = \alpha_0 (0.6)$$

A comparison of the third-order system with proportional controller, single-stage lead-controller and double-stage lead-controller is significant. The attenuations of the lead-controllers are made equal in order to provide a basis of comparison. The improvement obtainable with the double-stage lead-controller is marked.

	Third-Order Servo with Proportional Controller	Third-Order Servo with Basic Lead-Controller	Third-Order Servo with Double-Stage Lead-Controller
ζ_3	1.0	1.0	1.0
α	——	10	——
R_{d3}	——	0.2	——
α_0	——	------	$\sqrt{10}$
R'_{d3}	——	------	.316
R''_{d3}	——	------	.316
u_m	0.6	1.3	1.9
$\frac{K_p k_m}{J_d \omega_0^3}$	0.58	——	——
$\frac{k_d k_m}{J_d \omega_0^3}$	——	17.6	——
$\frac{k_{b1} k_{b2} k_m}{J_d \omega_0^3}$	——	——	19

TABLE 6

The double lead-controller is sometimes termed a first and second derivative controller. Although it is true that the output of such a controller is proportional approximately to the sum of the input and its first two derivatives, the controller in this example is not generally termed a first and second derivative controller, because the attenuation constant of each section is so small. In the past it has been generally considered that the value of a controller of this type with such a low attenuation constant was open to question; in fact, it has been considered that the value of using a second-derivative controller of any type was doubtful. The preceding example is ample proof of the utility of such a controller and the value of considering it not as a derivative controller but as a lead-controller.

The salient point in the preceding development is that the lead-controller was adjusted to "match" the servo in which it was incorporated. The transfer-function of the basic system was factored into two parts and each section of the lead-controller adjusted to compensate for one part. The natural frequency of the servo was thereby increased by a factor proportional to the square-root of the attenuation constant of the controller. This is an example of an important design principle that is further developed in following sections.

Third-Order Transfer-Function Servo with Compound Lead-Controller; Case 2: $\zeta_3 > 1.0$

For those cases in which the damping ratio, ζ_3, of the third-order system is greater than unity, the principle introduced in the preceding section of matching the controller to the system leads to a variation in the controller design, although the same controller circuit, figure 66, is usable. The transfer-function of the system to be considered is given by Equation (225). If the damping ratio, ζ_3 , is larger than unity, part of the expression can be factored in a fashion similar to the case previously considered.

$$K_T G_T(s') = \frac{k_{b1}k_{b2}k_m}{J_d a_{d1} a_{d2} \omega_0^3} \left[\frac{1}{s'(s' + \zeta_3 + \sqrt{\zeta_3^2 - 1})\,(s' + \zeta_3 - \sqrt{\zeta_3^2 - 1})} \right] \times$$

$$\frac{1 + R'_{d3}\alpha_{d1}s'}{1 + R'_{d3}\,s'} \times \frac{1 + R''_{d3}\,\alpha_{d2}s'}{1 + R''_{d3}\,s'} \tag{231}$$

The total attenuation of the controller is to be kept constant: therefore, let

$$\alpha_{d1}\alpha_{d2} = \alpha_0^2 \tag{232}$$

$$K_T G_T(s') = \frac{k_{b1}k_{b2}k_m}{J_d \alpha_0^2 \omega_0^3} \left[\frac{1}{s'(s' + \zeta_3 + \sqrt{\zeta_3^2 - 1})(s' + \zeta_3 - \sqrt{\zeta_3^2 - 1})} \right] \times$$

$$\frac{s' \quad \dfrac{1}{R'_{d3}\alpha_{d1}}}{\dfrac{s'}{\alpha_d} + \dfrac{1}{R'_{d3}\alpha_{d1}}} \times \frac{s' + \dfrac{1}{R''_{d3}\,\alpha_{d2}}}{\dfrac{s'}{\alpha_{d2}} + \dfrac{1}{R''_{d3}\,\alpha_{d2}}} \tag{233}$$

If

$$\frac{1}{R'_{d3}\,\alpha_{d1}} = \zeta_3 + \sqrt{\zeta_3^2 - 1} \tag{234}$$

$$\frac{1}{R''_{d3}\,\alpha_{d2}} = \zeta_3 - \sqrt{\zeta_3^2 - 1} \tag{235}$$

Equation (233) becomes

$$K_T G_T(s') = \frac{k_{b1} k_{b2} k_m}{J_d \alpha_0^2 \omega_0^3} \quad \frac{1}{s' \left(\frac{s'}{\alpha_{d1}} + \frac{1}{R'_{d3} \alpha_{d1}} \right) \left(\frac{s'}{\alpha_{d2}} + \frac{1}{R''_{d3} \alpha_{d2}} \right)} \tag{236}$$

If α_{d1} and α_{d2} were set equal to one another, the form of Equation (236) would lead to a conclusion similar to that of the preceding section: namely, that the response of the resultant system would be similar in form to that of the original system, but the natural frequency would by increased by a factor of α_0. However, even more improvement is obtainable. Equation (236) can be written in the following way if use is made of Equation (232).

$$K_T G_T(s') = \frac{k_{b1} k_{b2} k_m}{J_d \alpha_0^2 \omega_0^3} \quad \frac{1}{\frac{s'}{\alpha_0} \left(\frac{s'}{\alpha_0} + \frac{1}{R'_{d3} \alpha_0} \right) \left(\frac{s'}{\alpha_0} + \frac{1}{R''_{d3} \alpha_0} \right)} \tag{237}$$

Let $R'_{d3} = R''_{d3} = \frac{1}{\alpha_0}$ (238)

Then:

$$K_T G_T(s') = \frac{k_{b1} k_m k_{b2}}{J_d \alpha_0^3 \omega_0^3} \left[\frac{1}{\frac{s'}{\alpha_0} \left(\frac{s'}{\alpha_0} + 1 \right) \left(\frac{s'}{\alpha_0} + 1 \right)} \right] \tag{239}$$

A comparison of the bracketed portions of Equation (231) and (239) reveals a very interesting fact. Whereas the original system had a damping ratio greater than unity, the compensated system has a damping ratio equal to unity and in addition a resonant frequency equal to α_0 times the resonant frequency of a system <u>identical with the original system but with a unity damping ratio</u>. A comparison of the loci of figure 31 and the summary of Table 1 shows that a third-order system with a damping ratio of unity is superior to one in which the damping ratio is greater than unity. The proper matching of the double-stage lead-controller to the third-order servo has in essence decreased the damping ratio to unity and increased the resonant frequency of that system by a factor, α_0, both changes of which are desirable. A summary of the proper controller adjustments follows.

$$R'_{d3} = R''_{d3} = \frac{1}{\alpha_0} \tag{238}$$

$$\alpha_{d2} = \alpha_0 \left(\zeta_3 + \sqrt{\zeta_3^2 - 1} \right) \tag{240}$$

$$\alpha_{d1} = \frac{\alpha_0}{\zeta_3 + \sqrt{\zeta_3^2 - 1}} \tag{241}$$

$$\alpha_0 > \zeta_3 + \sqrt{\zeta_3^2 - 1}$$

Equation (241) requires that $\alpha_0 > \zeta_3 \pm \sqrt{\zeta_3^2 - 1}$ since α_{d1} must be larger than unity. If α_0 cannot be made larger than $\zeta_3 + \sqrt{\zeta_3^2 - 1}$, it is better to employ the basic lead-controller rather than a compound lead-controller.

Physically, the compensation process that has been employed can be described as follows: the third-order transfer-function has been divided into two factors with the form $\frac{1}{(s + c_1)}$.

This type of factor is known as a phase-lag factor. The double-stage lead-controller has been so adjusted that each stage tends to compensate for one of the phase-lag factors. The phase-lag factor with the small root $(s' + \zeta_3 - \sqrt{\zeta_3^2 - 1})$ requires more compensation than the other, so the attenuation constants of the two stages of the lead-controller have been so adjusted that the phase-lag factor requiring more compensation receives it. Since the attenuation constants have a limited variation if their product is fixed, the adjustment of the controller has been carried out in such a manner that maximum servo response is achieved for the available attenuation. However, if one phase-lag factor has a sufficiently small root, better results are achieved if all the attenuation available is applied to that factor, and the other factor neglected. This is the case when $\alpha_0 \leq \zeta_3 + \sqrt{\zeta_3^2 - 1}$.

A particular example in which the damping ratio, ζ_3, is equal to 1.41 has been calculated and the results are summarized in Table 7.

	Third-Order Servo with Proportional Controller	Third-Order Servo with Double-Stage Lead-Controller
ζ_3	1.414	1.414
α_0^2	—	10
R'_{d3}	—	.316
R''_{d3}	—	.316
α_{d1}	—	1.31
α_{d2}	—	7.63
$\frac{k_p k_m}{J_d \omega_0^3}$	0.50	—
$\frac{k_{b1} k_{b2} k_m}{J_d \omega_0^3}$	—	19.0
u_m	0.40	1.9

TABLE 7

Third-Order Transfer-Function Servo with Compound Lead-Controller; Case 3: $\zeta_3 < 1.0$

When the design of a compensating-controller is undertaken for a third-order servo in which the damping ratio, ζ_3 , is less than unity, it is found that the controller of figure 66, and the procedure hitherto employed, is not directly applicable. The reason is that the factors of the third-order transfer-function have complex roots if $\zeta_3 < 1.0$, while the factors of a cascaded lead-controller transfer-function have real roots only. However, controllers are realizable whose transfer-function will factor into complex portions and several are considered in detail in this paper. One type is developed in this section in order to illustrate the design principles involved when compensating a servo of such nature that its transfer-function factors into complex terms.

The schematic diagram of the controller that is to be applied to the problem is illustrated in figure 70. The impedance of the input and the resistances, r_1 and r_2 are considered to be

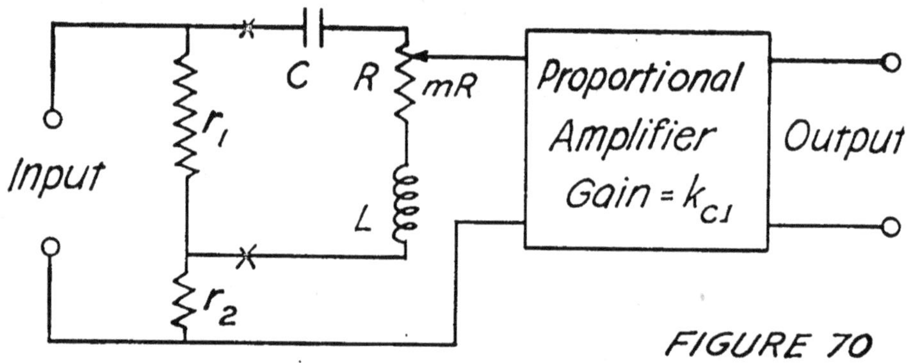

FIGURE 70

sufficiently small that the resistance looking toward the input from terminals xx is negligible. Not all the voltage developed across the resistance R is fed into the amplifier but only a fraction of it, denoted by "m". The transfer-function of this controller is denoted $K_{dc}G_{dc}(s)$ and its expression is

$$K_{dc}G_{dc}(s) = k_{c1}\left[\frac{r_2}{r_1 + r_2} + \frac{r_1}{r_1 + r_2}\frac{Ls + mR}{Ls + R + \frac{1}{Cs}}\right] \qquad (242)$$

Let

$$\frac{r_1 + r_2}{r_2} = \alpha_c^2 \qquad (243)$$

where α_c^2 is the attenuation constant of the controller.

$$K_{dc}G_{dc}(s) = k_{c1}\frac{LCs^2 + RC(m + \frac{1}{\alpha_c^2} - \frac{m}{\alpha_c^2})s + \frac{1}{\alpha_c^2}}{LCs^2 + RCs + 1} \qquad (244)$$

Let $\dfrac{1}{n} = m + \dfrac{1}{\alpha_c^2} - \dfrac{m}{\alpha_c^2}$ $\qquad (245)$

Equation (244) can be written

$$K_{dc}G_{dc}(s) = k_{c1} \frac{s^2 + \frac{R}{nL} s + \frac{1}{a_c^2 LC}}{s^2 + \frac{R}{L} s + \frac{1}{LC}} \tag{246}$$

Let $$\frac{1}{a_c^2 LC} = \omega_c^2 \tag{247}$$

$$\frac{R}{nL} = 2\zeta_c\omega_c \tag{248}$$

Then,

$$K_{dc}G_{dc}(s) = k_{c1} \frac{s^2 + 2\zeta_c\omega_c s + \omega_c^2}{s^2 + 2\zeta_c n\omega_c s + a_c^2\omega_c^2} \tag{249}$$

The form of the third-order transfer-function that will be employed in this derivation is that given by Equation (117a).

$$K_3G_3(s) = \frac{k_p k_m}{J_d\omega_0^3} \frac{1}{s'(s'^2 + 2\zeta_3 s' + 1)} \tag{117a}$$

The complex variable s' is a transformed variable:

$$\frac{s}{\omega_0} = s' \tag{116}$$

In terms of the same complex variable, Equation(249) becomes

$$K_{dc}G_{dc}(s') = k_{c1} \frac{s'^2 + 2\zeta_c \frac{\omega_c}{\omega_0} s' + \frac{\omega_c^2}{\omega_0^2}}{s'^2 + 2\zeta_c n \frac{\omega_c}{\omega_0} s' + a_c^2 \frac{\omega_c^2}{\omega_0^2}} \tag{250}$$

The transfer-function, $K_TG_T(s')$, of the combined system is obtained by replacing k_p, the gain factor of the proportional amplifier in Equation (117a) by Equation (249), the expression for the transfer-function of the compensating controller.

$$K_TG_T(s') = \frac{k_{c1}k_m}{J_d\omega_0^3} \frac{1}{s'(s'^2 + 2\zeta_3 s' + 1)} \frac{s'^2 + 2\zeta_c \frac{\omega_c}{\omega_0} s' + \frac{\omega_c^2}{\omega_0^2}}{s'^2 + 2\zeta_c n \frac{\omega_c}{\omega_0} s' + a_c^2 \frac{\omega_c^2}{\omega_0^2}} \tag{251}$$

If $$\omega_c = \omega_0 \tag{252}$$

$$\zeta_c = \zeta_3 \tag{253}$$

Equation (251) becomes:

$$K_T G_T(s') = \frac{k_{c1} k_{m3}}{J_d \omega_0^3} \quad \frac{1}{s'(s'^2 + \frac{2\zeta_3 n}{\alpha_c} \alpha_c s' + \alpha_c^2)} \tag{254}$$

Let

$$\frac{s'}{\alpha_c} = s'' \tag{255}$$

$$\frac{\zeta_3 n}{\alpha_c} = \zeta_T \tag{256}$$

$$K_T G_T(s'') = \frac{k_{c1} k_m}{J_d \omega_0^3 \alpha_c^3} \quad \frac{1}{s''(s''^2 + 2\zeta_T s'' + 1)} \tag{257}$$

A comparison of Equations (257) and (117a) reveals that the application of the controller of figure 70 has resulted in a new system, also of third-order, but whose natural frequency has been increased by the factor α_c, and whose damping ratio, ζ_T, is adjustable within wide limits. The factor n (see Equation (256)) can vary from unity to a_c^2 (since in Equation (245) m can vary from unity to zero) and thus result in a variation of the damping ratio, ζ_T, from $\frac{\zeta_3}{\alpha_c}$ to $\zeta_3 \alpha_c$. It is seen at once, therefore, that this type of controller is most flexible in its application; its use permits the correct compensation of a third-order system for any value of damping ratio within the range set by the attenuation factor of the controller. The factor m cannot be made zero in a physical system, because of the presence of resistance in a physical inductor. The value of m can generally be made sufficiently small, however, that this is not a basic limitation.

This type of lead-controller is termed a "matching" lead-controller in this paper, because of its flexibility of adjustment and ability to "match" a third-order function whose factors are either real and unequal, real and equal, or complex. Becuase of the significance of the results of the above design, the equations of importance in the synthesis and performance of this controller are grouped below:

Design Equations of the Matching Lead-Controller

$$\alpha_c^2 = \text{attenuation constant of controller} \tag{243}$$

$$\omega_c = \frac{1}{\alpha_c^2 LC} \quad \text{(set equal to } \omega_0\text{)} \tag{247}$$

$$\zeta_c = \frac{1}{2\omega_c} \quad \frac{R}{nL} \text{ (set equal to } \zeta_3\text{)} \tag{248}$$

$$m = \frac{\alpha_c^2 - n}{n(\alpha_c^2 - 1)} \quad \text{(damping ratio adjustment)} \tag{245a}$$

Resonant frequency of compensated system = u_{mT}

Resonant frequency of uncompensated system = u_{m3}

$$u_{mT} = \alpha_c u_{m3} \tag{258}$$

$$\zeta_T = \frac{n}{\alpha_c} \zeta_3 \tag{256}$$

$$K_T = \frac{k_{c1} k_m}{J_d \omega_0^3 \alpha_c^3} \text{ (Numerical value determined from Table 1 or Table 2).} \tag{259}$$

As an example, suppose that it is desired to synthesize a controller for a third-order system the damping ratio of which is 0.3. Assume that an improvement in the natural frequency of the system of about three is desired and an attenuation factor, α_c^2, of ten is chosen. Table 1 reveals that a damping ratio of approximately 0.7 is optimum for the third-order system;. therefore

$$\frac{n}{\alpha_c} = \frac{\zeta_T}{\zeta_3} = \frac{0.7}{0.3}$$

$$n = \sqrt{10} \left(\frac{0.7}{0.3} \right) = 7.4$$

$$m = \frac{\alpha_c^2 - n}{n(\alpha_c^2 - 1)} = \frac{10 - 7.4}{7.4\ (10 - 1)} \cong .04$$

$$\frac{1}{LC} = \alpha_c^2 \omega_c = 10\ \omega_0$$

$$\frac{R}{L} = 2\omega_c n \zeta_c = 2n\ \omega_0 \zeta_3$$

$$\frac{R}{L} = 2\ (7.4)(0.3)\ \omega_0 = 4.4\ \omega_0$$

$$u_{mT} = (3.16)(.65)\ \omega_0 \cong 2\omega_0$$

Synthesis of Lead-Controllers: Generalization

The preceding sections have dealt with the design of a servo controller for a particular type of servo. A simple example was chosen in order that system complexities would not obscure design principles; however, the example contained all the elements that must be considered in the design of complex systems, and the synthesis of the controller developed principles that can be followed in the synthesis of a lead-controller for a servo system of any complexity.

The frequency dependent portion of the transfer-function of a physical system can be written in the general form given by Equation (82):

$$G(s) = \frac{(s + b_1)(s + b_2)(s + b_3) \ldots}{s^n\ (s + c_1)(s + c_2)(s + c_3) \ldots} \tag{82}$$

The parenthesis terms in the denominator (other than the term s^n) are known as lag-factors; their number and the size of the terms c_1, c_2, etc., are an indication of the "slowness" of the servo system; it would be desirable if they were absent, but the fact that the servo is a physical system demands their presence, and it is the problem of the engineer to design compensating devices that reduce their effect.

The basic portions of the transfer-functions of many physical servo systems do not possess numerator terms such as $(s + b_1)$, $(s + b_2)$ $(s + b_3)$...in Equation (82). When these terms appear, they are often caused by compensating circuits introduced into the system and not by the elementary form of the servo. If these terms are present, the constants b_1, b_2, ... are positive or have positive real parts, provided the transfer-function corresponds to a "minimum phase system," as defined by Bode.[4] If certain of the terms, b_1, b_2, etc., are negative or have negative real parts, they can be removed by appropriate system design. Positive values of b_1, b_2, etc., do not contribute to the "slowness" of the servomechanism, since they may be considered as lead-producing factors.

In practically every case, therefore, it is the terms $(s + c_1)$, $(s + c_2)$, $(s + c_3)$ that require compensation if the speed of response of the system is to be augmented. The development of the preceding portion of this chapter is a direct guide to the design of compensating circuits and the prediction of the resultant performance of the system. If the number of lag terms is even, a cascaded series of controllers of the type of figure 70 can be designed, in which the number of sections is equal to one-half the number of lag-factors; or if the number of lag-factors is odd, a cascade combination of controllers of types illustrated by figure 70 and figure 58 can be designed. In general, the basic lead-controllers cascaded in multiple will compensate for those lag-factors that have real roots; those that do not have real roots occur in conjugate pairs and require compensating networks of the type of figure 70. In every case, however, a combination of these two types is sufficient.

In every practical design the attenuation constant of the compensating system is limited; therefore, one problem is to so allocate that value of attenuation among the various compensating stages that maximum servo improvement is secured. The allocation of attenuation is straightforward. Obviously those lag-factors with smaller roots require more compensation, and therefore, compensating controllers with larger attenuation constants than the lag-factors with large roots. Thus, if the transfer-function of the basic part of the servo is

$$KG(s) = \frac{1}{(s^2 + 2\zeta_1\omega_1 s + \omega_1^2)(s^2 + 2\zeta_2\omega_2 s + \omega_2^2)\ldots(s + c_1)(s + c_2)\cdots} \tag{260}$$

while the attenuation constants assigned to the matching lead-controllers of the type of figure 70 are α_1^2, α_2^2, α_3^2 ..., and to the basic lead-controllers are α_{d1} , α_{d2}, α_{d3}, then values are assigned to the attenuation constants according to relations (261) and (262). The term α_T is the product of all the attenuation constants, and has a value as large as necessary to secure adequate servo performance.

$$a_1\omega_1 = a_2\omega_2 = a_3\omega_3 \; \cdots = a_{d1}c_1 = a_{d2}c_2 \; \cdots \tag{261}$$

$$a_1^2 a_2^2 a_3^2 \cdots a_{d1} a_{d2} a_{d3} \cdots = a_T \tag{262}$$

The relations (261) and (262) are based on the premise that optimum servo response is secured provided the phase lags introduced by each of the lag-factors of (260) are approximately equal. In particular the compensating circuits are so adjusted that the transfer-function $K_T G_T(s)$, of the combined system is

$$K_T G_T(s) = \frac{1}{(s^2 + 2\zeta_a\omega_a s + \omega_a^2)(s^2 + 2\zeta_b\omega_b s + \omega_b^2)\cdots(s + c_a)(s + c_b)\cdots} \tag{263}$$

in which

$$\omega_a = \omega_b = \cdots = c_a = c_b = \cdots \tag{264}$$

Such an adjustment approaches the optimum. It is difficult to predict accurately the optimum values of the damping ratios ζ_a , ζ_b , ζ_c , ...although their values probably should be within the range 0.5 to 1.0. A study of the loci of Equation (263) for a particular system enables a more accurate prediction to be made of the correct values of the damping ratios.

CHAPTER VII

APPROXIMATION CRITERIA

The proper design of compensating circuits and the adjustment of servo parameters to achieve optimum servo operation requires that the servo system be characterized by a suitable mathematical relationship. The approach to the servo design problem utilized in this paper depends upon expressing in terms of frequency the relation between the servo output, $\theta_o(j\omega)$, and the servo controller input, $E(j\omega)$. This function has been termed the transfer-function of the servo and its study has yielded design criteria and procedures. One of the first steps, therefore, in the study of a new servo system and application of the theory previously developed is to obtain a functional relationship that adequately specifies the physical system without being so complex that the labor of analysis and computation is increased unduly. The importance of correctly characterizing the physical servomechanism is stressed in the early part of the preceding chapter. Criteria that will serve as guides in deciding whether or not certain factors can be neglected are very useful in simplifying the form of the mathematical expression for the transfer-function of a physical system. It is thus important to develop criteria that will serve as guides in deciding whether or not certain factors can be neglected in system analysis.

An example that introduces the principles involved in making approximations is the servo system illustrated in figure 71.

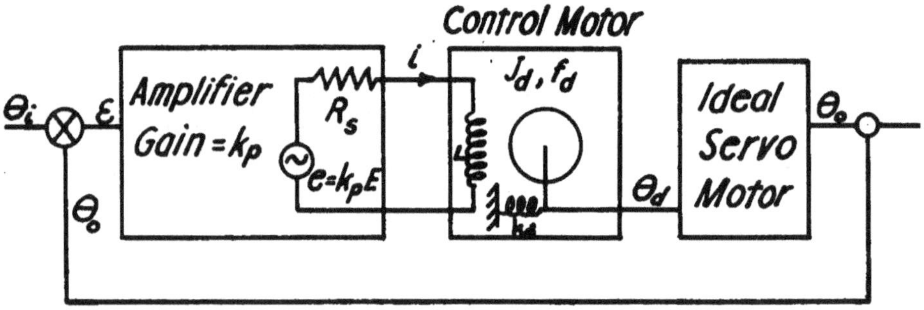

FIGURE 71

This system is very similar to the third-order transfer-function servo illustrated by figure 29; the difference is that the torque of the control motor is proportional to the current, i, instead of directly proportional to the error, ε. The current, i, is driven through a resistance-inductance circuit by an electromotive force, e, which in turn is proportional to the error, ε. The problem is to determine under what conditions the servo system of figure 71, which is a step closer to an actual physical system, can be analyzed as though it were the third-order system of figure 29. In other words, the problem is to determine the conditions under which the inductance, L, of the control-motor can be neglected without resulting in excessive errors.

The transfer-function of the system of figure 71 is given by Equation (265).

$$K_4G_4(s) = \frac{k_pk_mk_f}{J_dR}\;\frac{1}{(1 + \frac{L}{R}\, s)s(s^2 + 2\zeta_3\omega_0 s + \omega_0^2)} \qquad (265)$$

in which

L = inductance of control motor winding,

R = series resistance of circuit of control motor winding, (includes winding resistance and source resistance),

k_f= conversion factor between control motor torque and current, i; (equal to torque per ampere),

and the other terms have been defined on page 55.

Let the time constant, $\frac{L}{R}$, be denoted by τ_R,

$$\tau_R = \frac{L}{R} \qquad (266)$$

and make the transformation

$$s' = \frac{s}{\omega_0} \qquad (116)$$

Equation (265) becomes

$$K_4G_4(s') = \frac{k_pk_mk_f}{J_dR\omega_0^3}\left[\frac{1}{(1 + \tau_R\omega_0 s')}\right]\;\frac{1}{s'(s'^2 + 2\zeta_3 s' + 1)} \qquad (267)$$

The transfer-function, Equation (267), is a product of two functions: one that is the familiar third-order function and a second portion, in brackets, that is a function of the inductance-resistance circuit. The product $\tau_R\omega_0$ is a non-dimensionalized quantity that is a function of the inductance-resistance ratio of the circuit and the characteristic frequency, ω_0. The problem is to determine those values of $\tau_R\omega_0$ which allow the system to be considered a third-order system and permit the bracketed portion of Equation (267) to be omitted. The problem is solved by plotting a series of transfer-loci of the function of Equation (267) for particular values of the product $\tau_R\omega_0$ and from the shapes of these loci determining those values of $\tau_R\omega_0$ that leave the third-order locus substantially unaffected.

Figure 72 comprises such a series of loci. A particular value of damping ratio (namely, $\zeta_3 = 0.5$) was chosen and curves plotted for values of $\tau_R\omega_0$ equal to zero, 0.1, 0.2, and 0.5. The locus for which $\tau_R\omega_0$ is equal to zero is, of course, the third-order transfer-function locus. The principal effect at low frequencies of incorporating the lag-factor, $\frac{1}{(1 + j\tau_R\omega_0 u)}$, in the transfer-function is to shift the locus in a clockwise direction; that is, at low frequencies the lag-factor introduces some phase-shift but little amplitude distortion. **This is to be expected since the amplitude change introduced by the lag-factor is proportional to** $\frac{1}{\sqrt{1 + \tau_R^2\omega_0^2 u^2}}$,

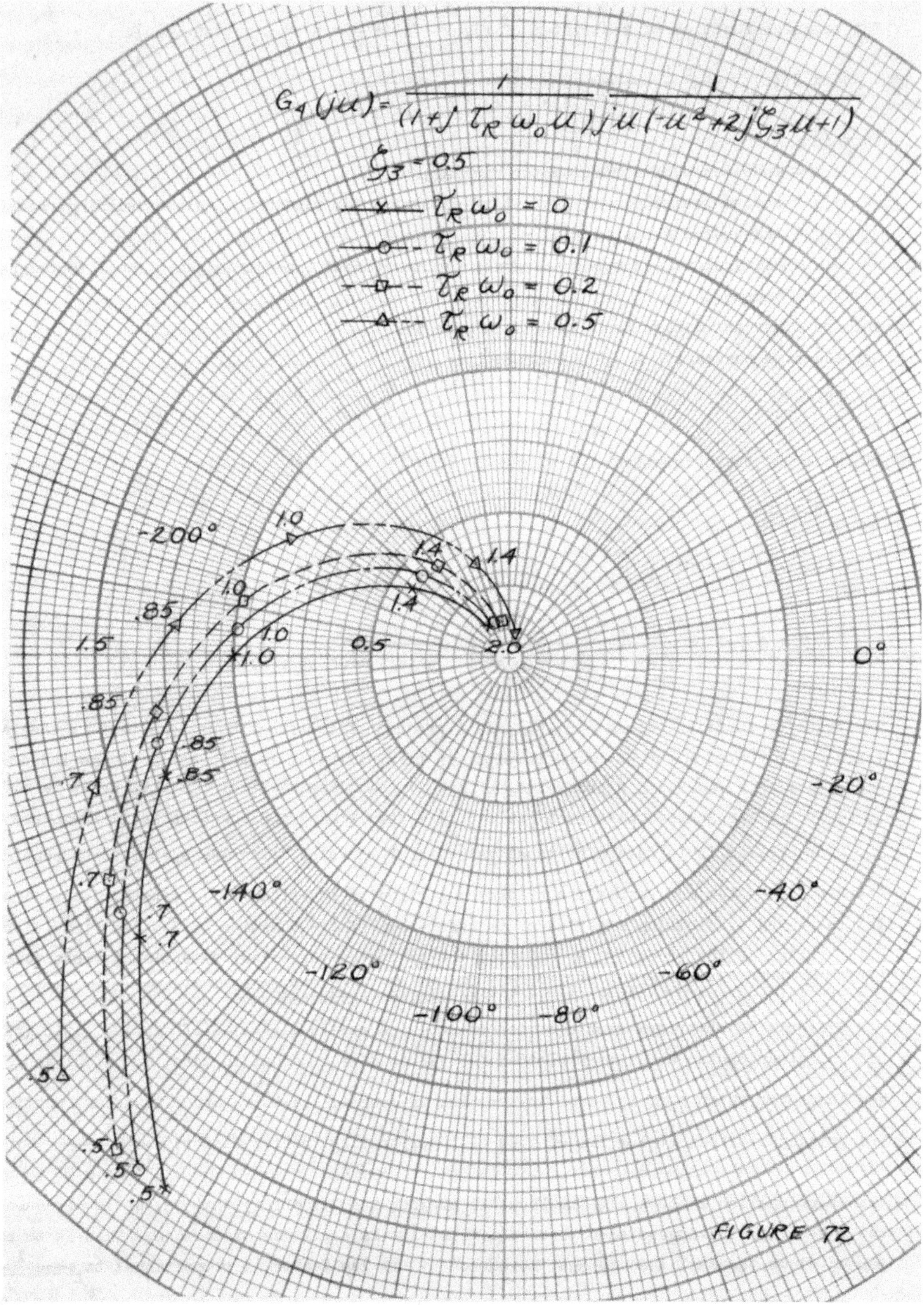

G4(jU) = 1/(1+j τR ω0 U) · 1/(jU(-U²+2jζ3U+1))
ζ3 = 0.5
τR ω0 = 0
τR ω0 = 0.1
τR ω0 = 0.2
τR ω0 = 0.5
-200°
1.5
0.5
2.0
0°
-20°
-40°
-60°
-80°
-100°
-120°
-140°
1.0
1.4
.85
.7
.5

FIGURE 72

which at frequencies for which $\tau_R\omega_0 u < 1.0$ is approximately equal to unity; on the other hand, the phase-shift introduced by the lag-factor is equal to $\tan^{-1}\tau_R\omega_0 u$ and can amount to several degrees even at frequencies for which $\tau_R\omega_0 u$ is considerably less than unity.

A measure of the effect of the lag-factor is obtained by determining the gain factor and resonant frequency of the system for each value of the product $\tau_R\omega_0$ for which a transfer-locus is plotted, and comparing the values obtained with those obtained for the third-order transfer-function. Table 8 summarizes this data.

TABLE 8

$\tau_R\omega_0$	K_4	u_m	u_m error	K_4 error
0	.46	.8	0	0
0.1	.43	.75	6.7%	7%
0.2	.41	.72	11%	12%
0.5	.39	.63	27%	18%

Thus if the lag-factor, $\dfrac{1}{1 + j\tau_R\omega_0 u}$, were neglected and $\tau_R\omega_0$ were equal to 0.2, the error in the prediction of the resonant frequency of the system would be the difference between 0.8 and 0.72, while the error in the prediction of the gain factor would be the difference between 0.46 and 0.41. The percentage errors are tabulated in Table 8. Thus, neglecting the lag-factor, if the product $\tau_R\omega_0$ is 0.2 causes a resonant-frequency error of 11 per cent and a gain-factor error of 12 per cent.

Errors of about 10 per cent in servo design generally are not excessive since the physical servo parameters are seldom known to an accuracy better than 10 per cent. On the other hand, the errors obtained if the product $\tau_R\omega_0$ is as large as 0.5 are greater than should be permitted in most system design. A safe conclusion, therefore, is that a lag-factor in which $\tau_R\omega_0$ is less than about 0.2 can be neglected; if the product $\tau_R\omega_0$ is greater than 0.2, it should be considered.

The influence a lag-factor exerts on a particular servo depends to a considerable extent upon the shape of the transfer-locus of the system. For example, if the transfer-locus in the vicinity of the negative real axis approximates a circle, the center of which lies at the origin, as illustrated in figure 73, curve A, the phase-shift introduced by the lag-factor will shift the locus along itself; that is, it will cause a frequency shift but only a very small change in the *shape* of the locus in this particular region with the result that the gain factor will be practically unaltered. On the other hand, if the locus tends to cross the negative real axis at an acute angle, as shown in figure 73, curve B, the phase-shift introduced by the lag-factor will affect the gain of the system more than the system resonant frequency. Because of these facts, it is difficult to set up an approximation criterion that can be relied upon without being too conservative. The best approach to the problem seems to be to base the criterion upon the phase-shift that the lag-factor introduces in the region of the resonant frequency of the basic system. For simple systems a safe rule is to permit a lag-factor to be disregarded

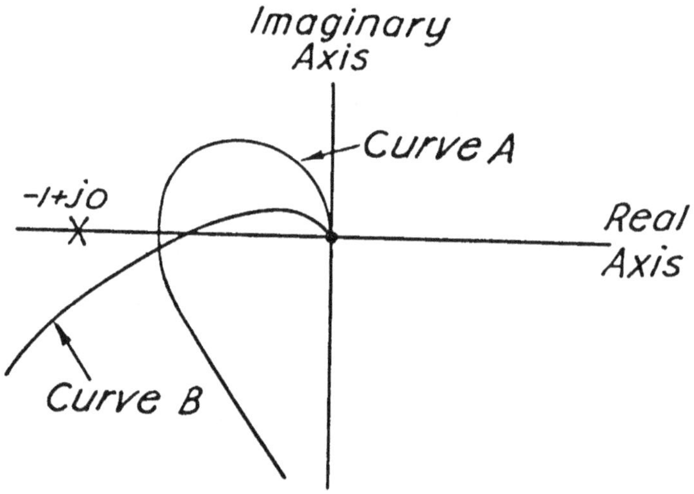

FIGURE 73

if the phase-shift it introduces at the resonant frequency of the system is not more than ten degrees. For more complex systems this figure may need revision. For simple systems (such as a third-order system) a ten-degree phase-shift produces an error in the gain factor and the resonant frequency of the system of about 10 per cent.

By expressing the above criterion in terms of the resonant frequency, account is taken of the fact that the permissible lag-factor varies with the natural frequency of the system. If several natural frequencies exist in the system, the phase-shift introduced by the lag-factor must not be greater than ten degrees at the highest natural frequency of importance in the system.

In the system previously considered and for which the loci of figure 72 were prepared, the resonant frequency is 0.8. If the phase of the lag-factor is limited to ten degrees at this frequency,

$$\text{arc}\left[1 + j\tau_R\omega_0(.8)\right] < 10^\circ$$

$$\tau_R\omega_0 < .22$$

which agrees with the previous conclusions.

When lead-controllers (lag-compensating controllers) are to be designed and incorporated in a system, an additional precaution must be taken in making approximations, as explained in the preceding chapter. In order to neglect a lag-factor in such a system, the phase-shift that it may introduce must be limited to about ten degrees at the resonant frequency of the compensated system—that is, the original system plus the lead-controller. This is entirely logical but it is frequently overlooked, and since the natural frequency of the compensated system may be several times that of the uncompensated system, the errors caused by neglecting lag-factors may be very large. Thus it may be permissible to neglect certain terms in the transfer-function if the system is to be operated without compensating circuits because the phase-shift these terms introduce at the natural frequency of the system is within ten degrees; however, if compensating circuits are employed and the natural frequency of the system is increased several-fold, the phase-shift these same transfer-function terms introduce at the new natural frequency is much larger and may be of such size that it is no longer negligible.

CHAPTER VIII

THEORY OF MINIMUM VELOCITY-ERROR SERVOMECHANISMS: PART 2

The theory underlying the design of minimum velocity-error systems has been presented in two parts in order that principles developed in the chapter on the theory of phase lead-controllers might be applied to the design of systems with velocity-error correction. The fundamentals of minimum velocity-error systems have been developed in Chapter V. The present chapter discusses the design principles of systems comprising both integral-controllers and lead-controllers, systems comprising more complex integral-controllers, and certain fundamental limitations upon the design of velocity-error compensating devices.

Systems Comprising Both Integral- and Lead-Controllers

Chapter V has dealt with the desirability of incorporating an integral-controller in a system in which it is necessary to minimize certain steady-state errors. Chapter VI has developed the theory of design and application of lead-controllers whose responsibility it is to minimize the transient errors in a servomechanism. Obviously, it may be advantageous to employ both integral- and lead-controllers in servomechanisms for certain applications, and it is helpful if certain factors are emphasized that are important in those applications requiring both types of compensating devices.

When lead-controllers are employed that completely match the transfer-function of the basic system, the design of the integral-controller is straightforward. For example, suppose that is is desired to incorporate a matching lead-controller and an integral-controller with the third-order transfer-function servo of figure 29. The preceding chapter has shown that if a lead-controller alone, of the type illustrated by figure 70, is incorporated with this system, a new system also of the third-order results, but with a transformed characteristic frequency ω_0 and a transformed damping ratio, ζ. The values of the new characteristic frequency and damping ratio may be chosen within a range limited only by the attenuation constant of the lead network.

In Chapter V, the optimum adjustments are deduced for a third-order system with integral-controller alone. Specifically, it was found that optimum results are secured if ζ_3, the damping ratio of the third-order system is 0.4, if R_{13}, the transformed time constant of the integral-controller is six to eight, and if the gain factor, K_T, is 0.3 to 0.4. These results are summarized on page 75 together with references to the defining equations of the parameters ζ_3, R_{13}, and K_T. The important point to observe is that the determining parameters of the system are found in terms of ω_0, the characteristic frequency of the third-order transfer-function servo. Now, if both a lead-controller and an integral-controller are to be employed, the matching lead-controller should be so adjusted that the characteristic frequency of the system is as high as necessary and the damping ratio, ζ, is equal to 0.4. The gain factor, K_T, and the integral time constant, R_{13}, should then be made equal to approximately 0.4 and 8 respectively, with the actual physical constants of the controller calculated from the equations referred to on page 75 in which ω_0 is now the transformed characteristic frequency of the system.

Thus the design of such a system is divided into three parts: (1) selection of the optimum parameters of the system with integral-control alone; (2) the design of a lead-controller that transforms the characteristic frequency and the damping ratio of the system while preserving the general form of the transfer-function; (3) the design of an integral-controller for the transformed system. For a prescribed value of the constant, R_{13}, the RC product of the integral-controller is inversely proportional to the characteristic frequency of the basic system, (see Equation (144)), while the length of time the controller requires to compensate for a suddenly applied input velocity is directly proportional to the RC product of the controller. Therefore, the lead-controller reduces this time interval by the same factor by which it increases the characteristic frequency of the system.

Equation (257) is the transfer-function of a third-order system with a matching lead-controller. The parameters n and α_c, Equation (256), are so adjusted that the transformed damping ratio, ζ_T, is 0.4. Suppose that the integral-controller of figure 37 is incorporated in the system, and that the integral-controller is so adjusted that the feedback factor, h, is equal to unity. If no lead-controller is used, the time constant, R_{13}, is equal to $RC\omega_0$ and its optimum value is six to eight. With a lead-controller in the circuit, the transformed time constant, R''_{13}, is still approximately eight, but it is now defined by

$$R''_{13} = RC\alpha_c\omega_0 \tag{268}$$

The transformed gain factor, K_T'', is still approximately 0.3 to 0.4, but it is now defined by Equation (259).

$$K_T'' = \frac{k_{c1}k_m}{J_d\omega_0^3\alpha_c^3} \tag{259a}$$

A procedure analagous to that outlined above is pursued when lead- and integral-controllers are synthesized for more complex servo systems. These more complex systems may be characterized by several characteristic frequencies and damping ratios. However, the design of compensating controllers may be carried out in a manner similar to the preceding example: (1) A determination of the damping ratios and the relation existing among the several characteristic frequencies of the basic servo system that ensures optimum performance when the basic system with freely adjustable parameters, employs an integral-controller. The optimum time constant of the integral-controller is determined at the same time. (2) Design of a matching lead-controller that transforms the parameters of the original system to those optimum values found in step (1), and increases the characteristic frequencies of the system by a sufficient factor that the requirements of the application are met. (3) Final selection of the system gain factor and the time constant of the integral-controller to correspond with the transformed characteristic frequencies of the system comprising the original system and the matching lead-controller.

Step (1) of the above design procedure may be carried through by plotting the transfer-loci of the system with integral-controller alone and so adjusting the time constant of the integral-controller together with the damping ratios and the ratios of the characteristic frequencies of the basic system that an optimum form of transfer-locus results. This can be an arduous task if the system is complex and all forms of loci plotted. Thoughtful consideration of the problem and application of guides developed in earlier parts of this paper will greatly mitigate the actual labor involved, however.

Step (2), the design of the matching lead-controller has been adequately explained in Chapter VI. Step (3), the final selection of the integral-controller time constant consists solely of applying the criteria developed in step (1) to the transformed system.

For those cases where practical limitations demand that lead-controllers be employed that do not match the basic system, no such detailed procedure can be outlined for the design of the compensating circuits. It is possible, however, to set down a design technique that yields approximately correct results in a majority of cases. The method makes use of the fact that the phase-shift introduced by the integral-controller in the region of the first and predominant resonant frequency of the system should be about ten degrees if the integral-controller is neither to have an unstabilizing effect on the system nor an excessively long time constant. This criterion is discussed more fully on page 70. The effect of the integral-controller upon the system transfer-locus in the region of the resonant frequency of the system is slight if this adjustment is made. On the other hand, the lead-controller must be so designed that its effect upon the system transfer-locus in the vicinity of the resonant frequency is marked. Therefore, since the major effects of the lead-controller and the integral-controller necessarily occur in separate regions of the frequency spectrum, an approximately correct result is obtained if their design is considered separately. The following procedure yields approximately optimum results: (1) Design the lead-controller as if it alone were to be employed with the basic system. An approximate method of designing the lead-controller that has been explained on page 101 is so to choose the time constant of the lead-controller that the maximum phase-shift, ϕ_{dm}, of the lead-controller occurs at a frequency for which the phase of the basic servo is greater by ϕ_{dm} than the phase at the resonant frequency of the basic system. (2) So design the integral-controller that the phase-shift that it introduces is approximately ten degrees at the resonant frequency of the system comprising the lead-controller and the basic system.

The above design technique is correct if the attenuation factor of the lead-controller is limited. (See page 101). If the attenuation factor can be large, it becomes very advantageous to employ matching lead-controllers, and the procedure previously explained should be employed.

Example: Third-Order Transfer-Function Servo with Basic Lead-Controller and Integral-Controller

The third-order transfer-function servo is considered as an example of the procedure to be followed when both an integral-controller and a basic lead-controller are employed. The transfer-function of the third order system is:

$$K_3G_3(ju) = \frac{k_pk_m}{J_d\omega_0^3}\;\frac{1}{ju(-u^2 + 2j\zeta_3u + 1)} \qquad (118a)$$

the symbols of which are defined on page 55. The transfer-function of a basic lead-controller is

$$K_{da}G_{da}(ju) = \frac{k_d}{a_d} \frac{1 + ja_d R_{d3} u}{1 + jR_{d3} u} \tag{192}$$

the symbols of which are defined on page 95-96. Finally, the transfer-function of an under-conpensating integral-controller is

$$K_{iu}G_{iu}(ju) = \frac{1}{1-h} \frac{1 + jR_{i3}u}{1 + j\dfrac{R_{i3}u}{1-h}} \tag{154a}$$

in which

$$K_{iu} = \frac{1}{1-h} \tag{269}$$

and the other symbols are defined on page 78. The factor $\frac{1}{1-h}$ is the attenuation constant of the integral-controller. Let $\frac{1}{1-h}$ be made equal to ten, and likewise, let the attenuation constant of the lead-controller equal ten. In order to compare the results with previous calculations, let the damping ratio, ζ_3, of the third-order system equal unity. The transfer-function of the complete system is found by forming the product of Equations (118a), (192), and (154a).

Following the procedure previously outlined, the first step is to design the lead-controller as if it alone were to be employed in the system. If the design-guide described on page 101 is followed, the value of R_{d3} is found to be about 0.2. The transfer-locus corresponding to the third-order system and lead-controller with this value of R_{d3} is plotted in figure 69.

The resonant frequency of the third-order system with basic lead-controller is equal to 1.3. The next step in the system design is to impose on the transfer-function of the integral-controller the requirement that its phase-shift at a frequency u = 1.3 equals ten degrees. Therefore,

$$\text{arc}\left[G_{iu}(j1.3)\right] = \text{arc}\left[\frac{1 + jR_{i3}(1.3)}{1 + j10\,R_{i3}(1.3)}\right] = 10^\circ \tag{270}$$

The construction of figure 46 then yields,

$$\frac{1}{\dfrac{R_{i3}\,(1.3)}{h}} = \tan 10^\circ \tag{271}$$

and R_{i3} is calculated to be 3.9.

The loci of figure 74 have been prepared to check the approximate design indicated above. Values of the loci parameters are $\zeta_3 = 1.0$, $R_{d3} = 0.2$, $a_d = 10$, $h = 0.9$, and $R_{i3} = 3.0$, 4.0, 6.0, and infinity. An infinite R_{i3} corresponds to the system with the integral-controller absent—that is, a third-order system with a lead-controller only. The shapes of the loci indicate that while there is not a great deal of difference between the performance of the system with settings of R_{i3} equal to 3.0, 4.0, and 6.0, that a value of 4.0 is probably closest to the

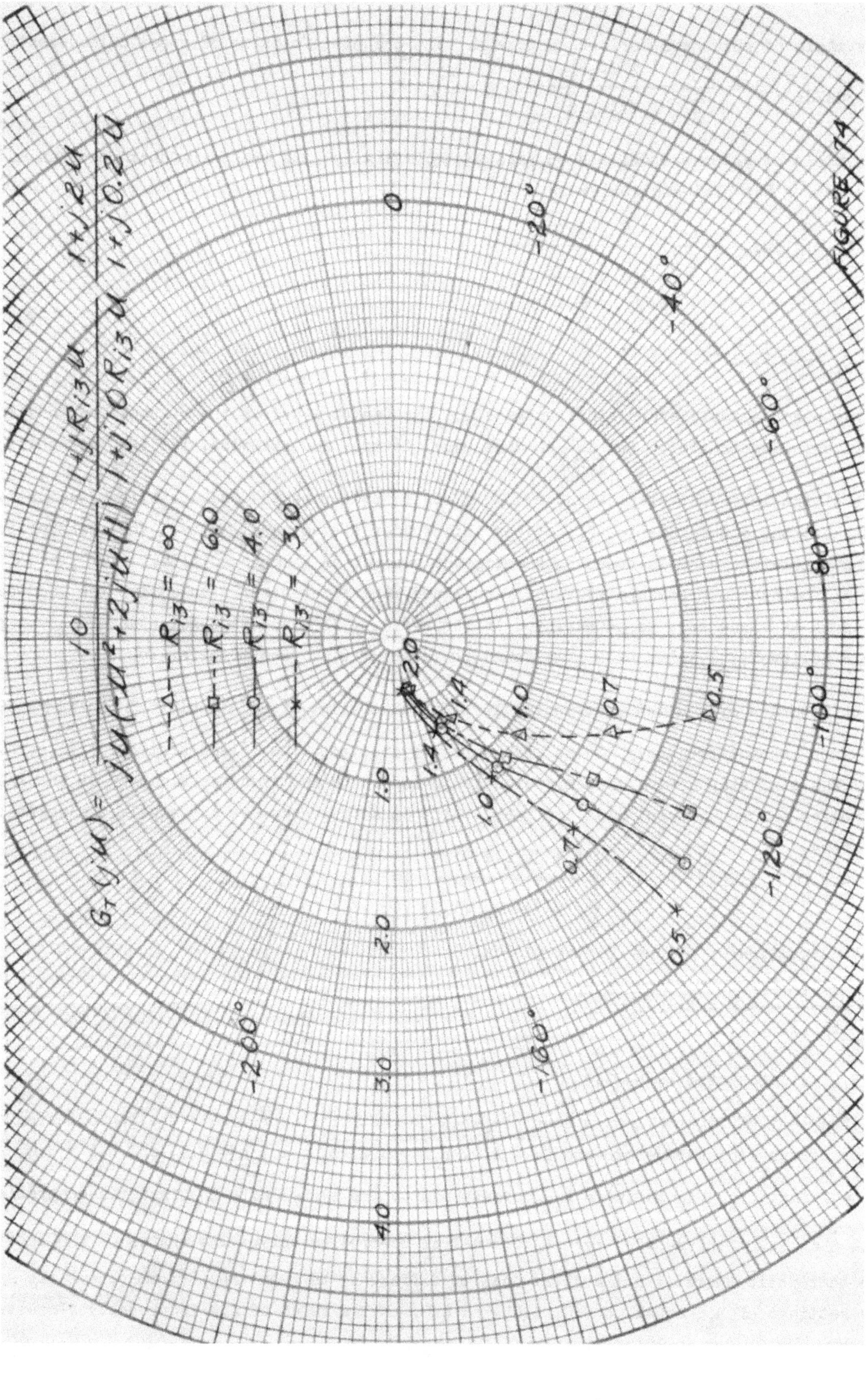
$G_T(jU) = \frac{10}{jU(-U^2+2jU+1)} \cdot \frac{1+jR_{13}U}{1+j10R_{13}U} \cdot \frac{1+j2U}{1+j0.2U}$
$R_{13} = \infty$
$R_{13} = 6.0$
$R_{13} = 4.0$
$R_{13} = 3.0$
0
-20°
-40°
-60°
-80°
-100°
-120°
-160°
-200°
1.0
2.0
3.0
4.0
0.5
0.7
1.0
1.4
2.0

FIGURE 74

optimum value.

Integral-Controller Design Limitations

The preceding section has shown how the incorporation of a lead-controller into a system with an integral-controller permits the RC product of the integral-controller to be reduced. The RC product of the integral-controller is a measure of the time interval required by the controller to correct a suddenly-applied velocity-error, and reducing this product is an important problem in many servo applications. Certain servo applications, however, require that the servomechanism distinguish between true datum which is an accurate indication of input and output positions, and false data caused by gear irregularities and other types of noise that tend to misdirect the servo controller. Since the frequency spectrum of the false data usually is higher than that of the true datum, acceptance or rejection can be effected upon the basis of the frequency of the information. In other words, the servo behaves like a low-pass filter, passing with a minimum of discrimination low-frequency signals and rejecting high-frequency signals. A servomechanism, by nature of its being a physical device, inherently possesses this characteristic to a certain extent, but certain applications demand a more pronounced low-pass characteristic. Therefore, certain filter networks are sometimes added to the controller to achieve this effect.

A moment's consideration reveals that, in general, lead-controllers should not be employed in applications in which the false data are particularly strong and have a frequency spectrum that lies close to the spectrum of the true datum. This is because the beneficial results secured by the use of lead-controllers arise from the fact that the lead-controller accentuates the response of the system to high frequencies, and therefore also accentuates false data signals that may be present. However, it may still be necessary to minimize the time required for the servo to compensate for a suddenly-applied input velocity, and if the use of lead-controllers is excluded by the presence of noise, it is nesessary to devise other methods for reducing the time constant of the integral-controller. The complete solution of this problem requires the design of special networks and their incorporation in an integral-controller, and therefore, requires extensive investigation and application of network theory. It is not the purpose of this paper to work out that solution but only to indicate the direction the investigation might take and to point out certain fundamental limitations that are encountered.

The developments in Chapter V have centered about three controllers: the first is a device the output of which is proportional both to the input and the true time integral of the input, and the second and third are devices that closely approximate the first device without yielding true integral response. The three devices have been formally termed an ideal integral-controller, an over-compensating integral-controller and an under-compensating integral-controller, although the general term "integral-controller" has been used informally to designate all three devices. It has been shown that the transfer-locus of an ideal integral-controller is a curve such as curve B of figure 40, and that the operation of the integral-controller depends upon the fact that it has very little effect upon the system at high frequencies, but provides very high gain at low frequencies and infinite gain at zero frequency. It has also been shown that proper adjustment of an integral-controller is secured by limiting to a small value the phase-shift it introduces into the system at the fundamental resonant frequency of the servomechanism. Once this controller is adjusted to provide a given phase-shift at a given frequency, its characteristic over the entire frequency spectrum is completely determined. (See Equation (139a).

It is possible to visualize the characteristic of an ideal velocity-error correcting controller. For frequencies in the neighborhood of the resonant frequency of the system, an ideal system would have unity amplitude response and a very small negative phase-shift. At frequencies substantially below the resonant frequency, the controller would have very large amplitude response and would by necessity possess a phase-shift in the neighborhood of ninety degrees. The shift from the characteristics of one region to those of the second should require a frequency band of minimum width.

The transfer-locus of such a hypothetical controller is illustrated by curve A, figure 75. Curve B is the locus of a third-order transfer-function, and curve C is the locus obtained by plotting the product of the transfer-functions corresponding to curves A and B.

Unfortunately, the forms that the transfer-locus of a physical device may take are limited and controlled by explicit laws. It is impossible to synthesize a physical controller that has both its amplitude and phase response prescribed unless attention is paid to the inter-relationship that exists between these two function. The problem that confronts the engineer in the design of a velocity-error compensating circuit is that of realizing a physical circuit with an amplitude characteristic having a sharp cut-off without a detrimental phase characteristic. It is always true that the more rapidly the attenuation of a physical network changes with frequency, the larger will be the phase-shift, associated with the network.

Bode[4] has shown that while there may be an infinite number of phase characteristics associated with a particular amplitude characteristic, only one of these is a <u>minimum</u> phase characteristic, and this minimum phase characterictic is completely determined by the given amplitude characteristic. Therefore, if a desirable amplitude characteristic is chosen, the minimum phase characteristic associated with this amplitude characteristic may be calculated; if this phase characteristic is unsatisfactory, the only alternative is to choose another and probably less favorable amplitude characteristic with the hope that an improved phase characteristic will result. The problem can be attacked not quite as blindly as the above would indicate. Bode has developed certain basic limitations upon, and interconnections between, amplitude and phase characteristics and certain short-cut methods for calculating one from the other.

<u>RLC Integral-Controller</u>

By way of illustrating the possibilities and limitations of other types of velocity-error correcting mechanisms, one such device is described. The network is an obvious one to investigate and was obtained by reasoning intuitively that such a network should increase the rate of change of attenuation with frequency over that obtainable with an ideal integral-controller.

Figure 37 is the general diagram of a controller which can be adjusted to have a pole at zero frequency. It has been shown that an RC network, such as illustrated in figure 39, inserted in the feedback path of the controller of figure 37, introduces a term in the controller output that is proportional to the integral of the controller input. With an RC feedback link the

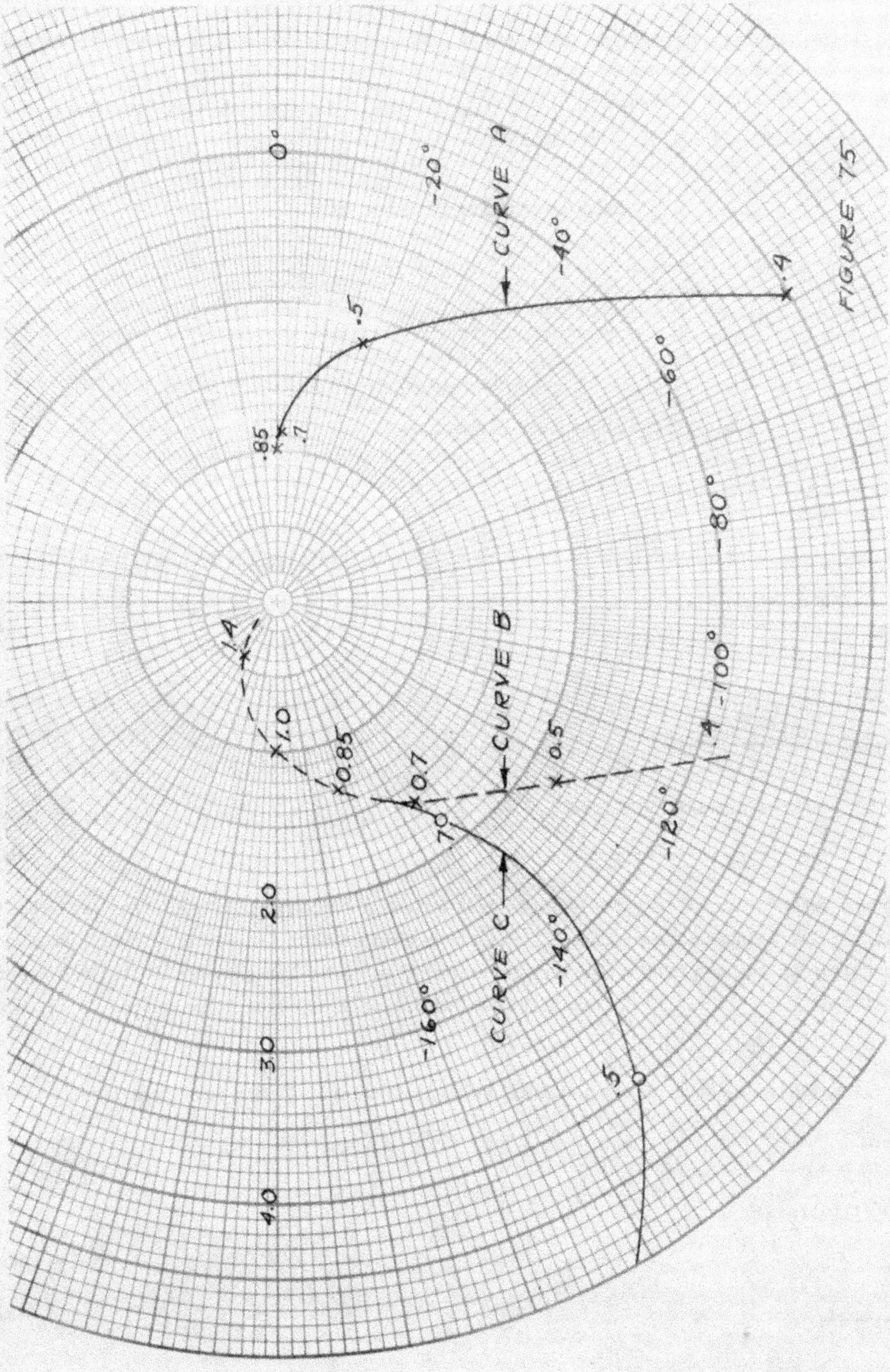
0°
-20°
-40°
-60°
-80°
-100°
-120°
-140°
-160°
CURVE A
CURVE B
CURVE C
2.0
3.0
4.0
.4
.5
.85
.7
1.4
1.0
0.85
0.7
0.5
.4
.7
.5

FIGURE 75

controller characteristic is specified by Equation (139a). Other types of circuits can be employed in the feedback link that have a sharper cut-off characteristic than the simple RC circuit in order to achieve an increase in the rate of change of attenuation with frequency. The RLC circuit illustrated by figure 76 is one such circuit.

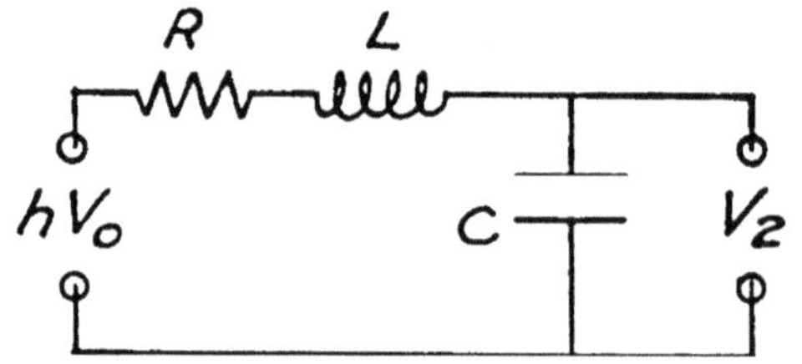

FIGURE 76

The transfer-function, L(s), of the feedback circuit is

$$L(s) = \frac{V_2(s)}{hV_0(s)} = \frac{1}{LCs^2 + RCs + 1} \tag{272}$$

The transfer-function of the controller is obtained by substituting Equation (272) into Equation (135):

$$K_iG_i = \frac{h(LCs^2 + RCs + 1)}{LCs^2 + RCs + 1 - h} \tag{273}$$

Equation (273) has a pole at s = 0 if h is equal to unity.

$$K_iG_i(s) = \frac{LCs^2 + RCs + 1}{s(LCs + RC)} \tag{274}$$

Equation (274) can be written as follows:

$$K_iG_i(s) = 1 + \frac{1}{RCs} - \frac{1}{RC(s + \frac{R}{L})} \tag{275}$$

The first two terms of Equation (275) are the same as those of Equation (139a), the transfer-function of an ideal integral-controller. Thus it is evident that the transfer-function of the above controller is the sum of the transfer-function of an ideal integral-controller and a perturbation factor, $-\frac{1}{RC(s + \frac{R}{L})}$. It is the purpose of the perturbation factor to improve the form of the transfer-locus of the ideal integral-controller. If the transformation

$$s' = \frac{s}{\omega_0} \tag{116}$$

and the substitutions

$$R_{13} = RC\omega_0 \tag{144}$$

$$\frac{L}{R} = \frac{R_{13}}{r\omega_0} \tag{276}$$

are made, Equation (274) becomes

$$K_1G_1(s') = \frac{s'^2 + \frac{r}{R_{13}} s' + \frac{r}{R_{13}^2}}{s'(s' + \frac{r}{R_{13}})} \tag{277}$$

In terms of the transformed frequency,

$$u = \frac{\omega}{\omega_0} \tag{119}$$

Equation (277) is

$$K_1G_1(ju) = \frac{-u^2 + j\frac{r}{R_{13}} u + \frac{r}{R_{13}^2}}{ju(ju + \frac{r}{R_{13}})} \tag{278}$$

The constant, R_{13}, is also the transformed time-constant of the ideal integral-controller. (See Equation (144), Page 72). The amplitude and phase characteristics of the RLC integral-controller are compared with those of the ideal integral-controller in figure 77. At high frequencies, $R_{13}u > 2.0$, the phase-shift of the RLC controller is considerably less than that of the integral-controller. On the other hand, at frequencies for which $R_{13}u < 1.0$, the phase-shift of the RLC controller is larger than that of the ideal integral-controller. There is not a great deal of difference between the amplitude characteristics of the controllers. Therefore, if the smaller phase-shift at high frequencies of the RLC controller permits a smaller time constant, R_{13}, to be employed with this controller, the gain of the RLC controller will be larger than that of the ideal, RC, integral-controller for the same low value of the frequency, u.

In order to illustrate the effect of the RLC integral-controller upon a servomechanism, the third-order transfer-function servo has been taken as an example. The value of r (See Equation (276)) has been set at 2.0, and the value of R_{13} varied. The results are plotted in figure 78. The transfer-locus of the same system with an ideal integral-controller is included in the figure for comparison purposes. The loci show that while the RLC integral-controller has an improved shape in the vicinity of the resonant frequency of the system, the larger phase-shift of the controller at lower frequencies causes the loci to bend back toward the negative real axis and determines the minimum value of the time-constant, R_{13}, that can be employed. The RLC circuit is an improvement over the RC circuit, however, in that the time constnat, R_{13}, can be reduced from a value of eight necessary when the RC circuit is employed to a value of six or possibly to as low a value as four. Thus it is evident that a system incorporating the RLC integral-controller would have better performance in that the length of time required to correct a suddenly-applied velocity-error would be reduced and the system error caused by sinusoidal inputs of low frequency would be diminished. The fundamental limitation on circuits of this type, however, is illustrated by the fact that the improvement in system performance is not nearly as large as might be expected from a casual investigation of the circuit.

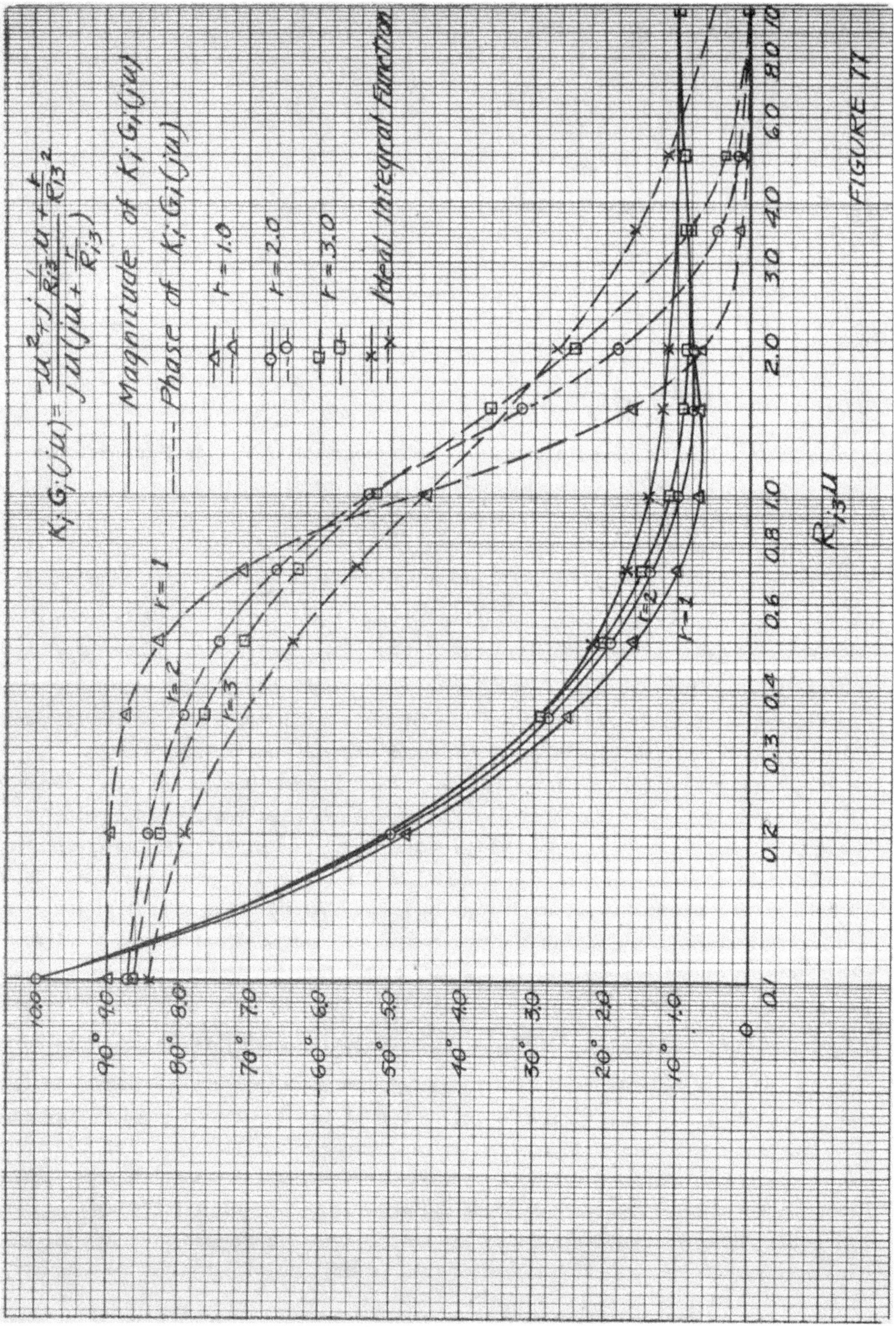

$$K_i G_i(ju) = \frac{-u^2 + j\frac{1}{R_{i3}}u + \frac{r}{R_{i3}^2}}{ju\left(ju + \frac{r}{R_{i3}}\right)}$$

FIGURE 77

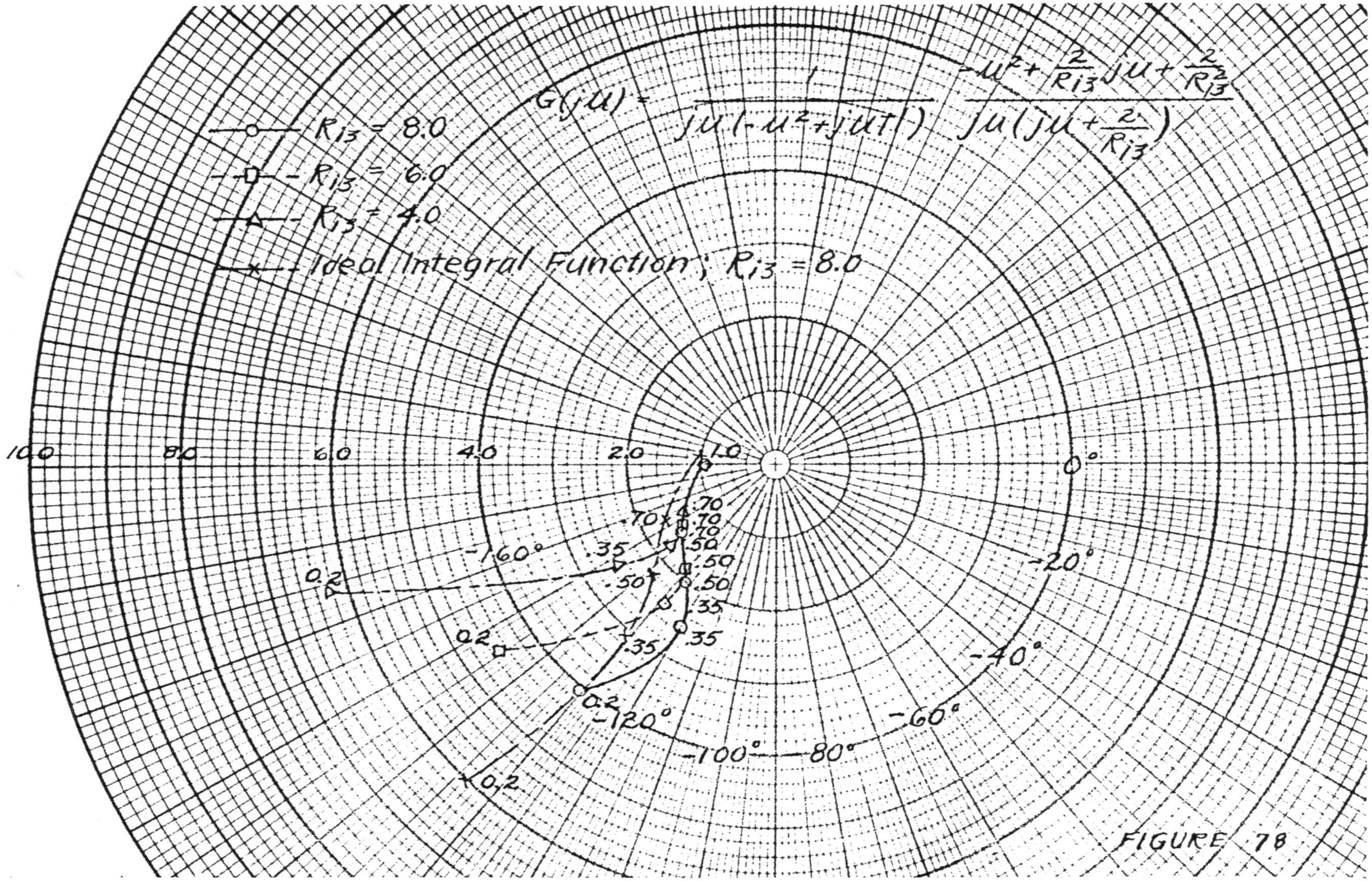

$G(ju) = \frac{1}{ju(1-u^2+juT1)} \cdot \frac{-u^2+\frac{2}{R_{13}}ju+\frac{2}{R_{13}^2}}{ju(ju+\frac{2}{R_{13}})}$
$R_{13} = 8.0$
$R_{13} = 6.0$
$R_{13} = 4.0$
Ideal Integral Function; $R_{13} = 8.0$
10.0
8.0
6.0
4.0
2.0
1.0
0°
-20°
-40°
-60°
-80°
-100°
-120°
-160°
.70
.50
.35
0.2

FIGURE 78

It is quite possible that more complex circuits would improve the performance of the controller still further. The best compensating circuit is not known, but probably can be found or approximated by the application of network theory with close attention paid to the relations between the minimum phase and amplitude functions as developed by Bode.

It should be pointed out here that these relations between phase and amplitude were discovered by N. Wiener and initially divulged by Y. W. Lee in a doctorate thesis done at M.I.T.[30]

CHAPTER IX

BASIC CONTROLLER NETWORKS

The preceding portion of this paper has dealt principally with the development of what has been termed compensating functions. The emphasis has been placed upon selecting the form and adjusting the parameters of these functions in such a manner that the steady-state and dynamic accuracies of a servomechanism may be improved to a maximum extent; not as much prominence has been given to the physical realization of these compensating functions although it has been shown by example that in each case the function could be physically realized. It is the purpose of this and subsequent chapters to develop and illustrate means of synthesizing physical devices that realize the characteristics of the compensating functions. No attempt is made to cover the synthesis problem in complete detail but the types of circuits described are sufficiently varied that the design and synthesis of similar circuits not actually mentioned should be straightforward. The intention of this and subsequent chapters is to develop methods of synthesis rather than to explore all the physical possibilities.

The present chapter first reviews the three general types of compensating functions that have been discussed in the preceding portion of the paper and second shows how a variety of simple types of circuits realize these general functions.

Chapters V and VIII of this paper have shown that the steady-state errors of a servomechanism are minimized by employing an integral-controller, the transfer-function of which can be written in the following general form.

$$K_iG_i(s) = k_i \frac{1 + \tau_i s}{1 + \alpha_i \tau_i s} , \qquad (279)$$

in which

$K_iG_i(s)$ = transfer-function of integral-controller

τ_i = time constant of integral-controller

α_i = attenuation factor of network providing the required functional relationship,

k_i = gain-factor of controller including the associated proportional amplifier.

The form of Equation (279) is familiar by now and can be compared with Equations (154) and (162) The ideal integral-controller is a special case of Equation (279).

Chapter VI has shown that the dynamic errors of a servomechanism can be minimized by correctly employing compensating controllers possessing the following two types of transfer-functions. The first is that of the basic lead-controller and has the general form given by Equation (192) and discussed on page 96.

$$K_{da}G_{da}(s) = \frac{k_d}{\alpha_d} \frac{1 + \tau_d \alpha_d s}{1 + \tau_d s} \qquad (192)$$

$K_{da}G_{da}(s)$ = transfer-function of the basic lead-controller

τ_d = time constant of the lead-controller

α_d = attenuation constant of the network providing the required functional relationship,

k_d = gain factor of controller including the associated proportional amplifier.

The second general type of lead-controller has been termed a matching lead-controller. Its transfer-function is given by Equation (149) and its characteristics are discussed on page 122.

$$K_{dc}G_{dc}(s) = k_{c1}\,\frac{s^2 + 2\zeta_c\omega_c s + \omega_c^2}{s^2 + 2\zeta_c n\omega_c s + \alpha_c^2\omega_c^2}\,, \tag{249}$$

in which

$K_{dc}G_{dc}(s)$ = transfer-function of the matching lead-controller.

α_c^2 = attenuation factor of the network providing the required functional relationship.

n = network constant determining the effect of the controller on the damping ratio of the system.

$\left.\begin{matrix}\omega_c\\ \zeta_c\end{matrix}\right\}$ = constants of the controller.

k_{c1} = gain factor of the controller, including the associated proportional amplifier.

It has been shown that the proper utilization of the above three functions, Equation (279), Equation (192), and Equation (249) permits the transient and steady-state errors of a servomechanism to be reduced by any required factor. These three functions, of course, are not the only ones that will provide compensation of servo errors, but they are the most simple ones, and simplicity is generally of importance in the production and maintenance of physical devices. A design procedure that has been found effective is to consider the above three functions as separate from any particular physical network and determine the optimum value of their constants for the conditions of a particular servomechanism design. After these constants have been determined (by application of the methods previously explained) a physical device can be selected that will properly realize the function and meet whatever other restrictions are imposed by the application. The problem of realization, now considered, is attacked from the standpoint of electrical networks, but the mechanical, hydraulic or other analogs of the electrical circuits should be obvious.

Under-Compensating Integral-Controller: Effect of Source and Load Resistances

A possible circuit for realizing the function of Equation (279) is illustrated in figure 54 and has been discussed in Chapter V. Its transfer-function is given by Equation (162):

$$K_1G_1(s) = k_a\,\frac{1 + R_2C_1s}{1 + (R_1 + R_2)C_1s} \tag{162}$$

If Equation (162) is compared with Equation (279), it is seen that the two are equivalent provided that

$$k_a = k_i \tag{280}$$

$$R_2C_1 = \tau_i \tag{281}$$

$$\frac{R_1 + R_2}{R_2} = a_i \tag{282}$$

Equation (162) is based upon the assumption that the source resistance and output resistance of the circuit of figure 54 are zero and infinite respectively. Every physical system with which the integral-controller network can be employed will possess a finite source resistance and a non-infinite load resistance. It is of interest, therefore, to determine the effect of these two quantities upon the transfer-function of the controller. However, figure 79 illustrates a circuit in which both source and load resistances are present.

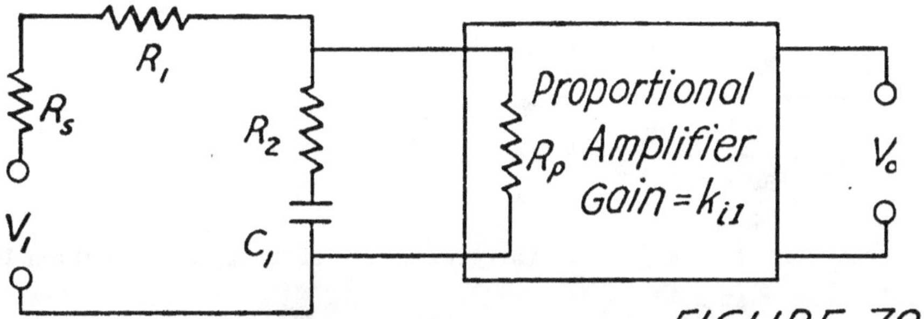

FIGURE 79

By application of Thevenin's theorem, the circuit can be transformed to that of figure 80.

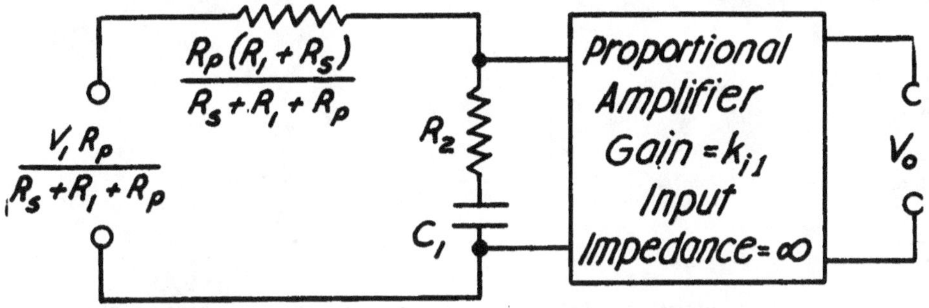

FIGURE 80

It is obvious that the circuit of figure 80 is equivalent to that of figure 54. Therefore, the circuit of figure 79 has a transfer-function with the form of Equation (279) provided that

$$k_{i1} = \frac{R_p + R_s + R_1}{R_p} k_i \tag{283}$$

$$R_2C_1 = \mathcal{T}_i \tag{284}$$

$$1 + \frac{R_p(R_1 + R_s)}{R_2\,(R_p + R_1 + R_s)} = a_i \tag{285}$$

Under-Compensating Integral-Controller: Inductive Circuit

The circuit of figure 54 has an equivalent that employs an inductance instead of a capacitance as the frequency-dependent element. The circuit is illustrated by figure 81. Its transfer-function is given by Equation (286).

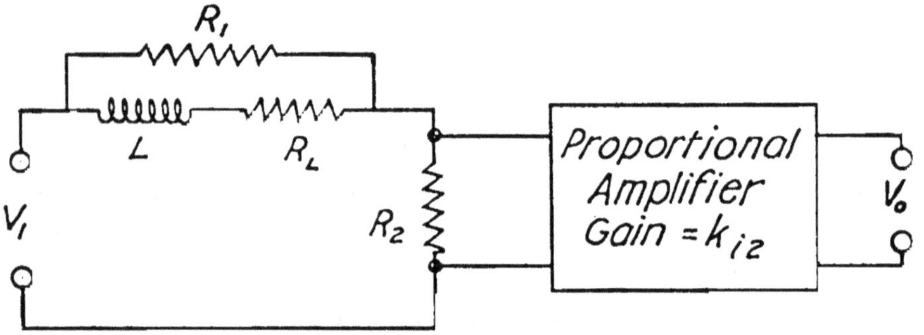

FIGURE 81

$$K_1G_1(s) = k_{12} \frac{R_2}{R_2 + \frac{R_1R_L}{R_1 + R_L}} \frac{1 + \frac{L}{R_1 + R_L} s}{1 + \frac{R_1 + R_2}{R_2(R_1 + R_L) + R_1R_L} Ls} \tag{286}$$

In order to be equivalent to the general integral-controller expression given by Equation (279), the following relations must hold:

$$\frac{R_2k_{12}}{R_2 + \frac{R_1R_L}{R_1 + R_L}} = k_1 \tag{287}$$

$$\frac{L}{R_1 + R_L} = \tau_1 \tag{288}$$

$$\frac{R_1 + R_2}{R_2 + \frac{R_1R_L}{R_1 + R_L}} = a_1 \tag{289}$$

The presence of source and output resistances can be handled in a manner similar to that employed in the previous example. The circuit of figure 81 is not as important practically as that of figure 54, because the value of the integral time constant, τ_1 , is limited to not more than one-tenth to one second by the size of the series resistance necessarily associated with every physical inductor.

Basic Lead-Controller: Effect of Source and Load Resistances

It has been previously demonstrated that Equation (192),

$$K_{da}G_{da}(s) = \frac{k_d}{\alpha_d}\,\frac{1 + \tau_d\alpha_d s}{1 + \tau_d\, s}\,, \tag{192}$$

the general form of the transfer-function of the basic lead-controller, is physically realized by the network illustrated by figure 58. The circuit relations, developed on page 95, are based upon the assumption that the source and load resistances of the circuit of figure 58 are zero and infinite respectively. In order to determine the effect upon the controller characterictic of a finite source resistance and a non-infinite load resistance, the circuit of figure 82 is examined. This circuit can be replaced by that of figure 83, which is equivalent in every respect.

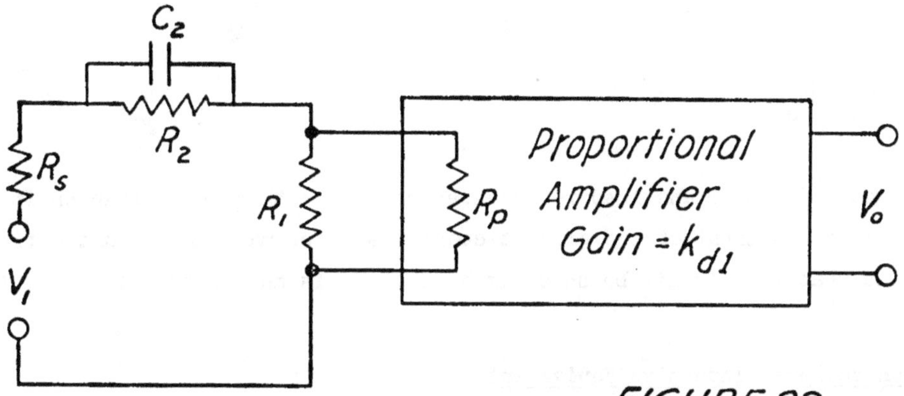

FIGURE 82

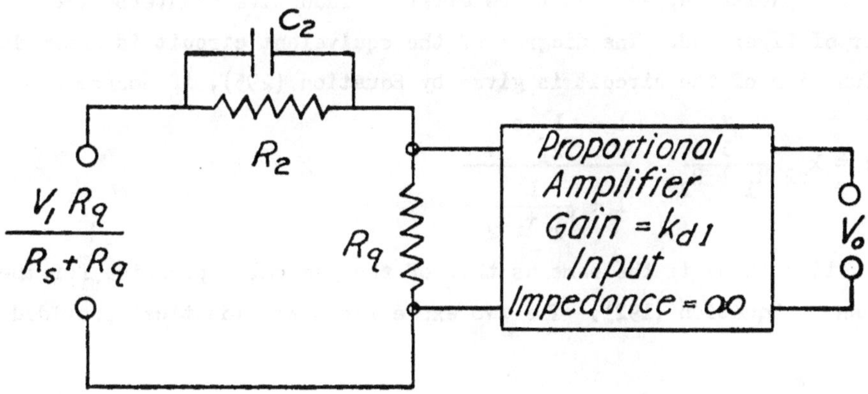

FIGURE 83

The resistance, R_q, is equal to the parallel resistance of R_1 and R_p.

$$R_q = \frac{R_1 R_p}{R_1 + R_p} \tag{290}$$

The transfer-function of the circuit is

$$K_{da}G_{da}(s) = \frac{R_q k_{d1}}{R_s + R_q} \quad \frac{R_q}{R_q + R_2} \quad \frac{1 + R_2 C_2 s}{1 + \dfrac{R_2 R_q}{R_q + R_2} C_2 s} \qquad (291)$$

The function (291) is equivalent to the general form for the basic lead-controller provided the follwing relations obtain:

$$\frac{R_q k_{d1}}{R_s + R_q} = k_d \qquad (292)$$

$$\frac{R_q + R_2}{R_q} = \alpha_d \qquad (293)$$

$$\frac{R_2 R_q}{R_2 + R_q} C_2 = \tau_d \qquad (294)$$

Thus the presence of source and load resistances require nothing more than an increase in amplifier gain and a readjustment of the circuit constants. For maximum circuit efficiency, however, it is important that the circuit be so designed that R_q is much larger than R_s, the source resistance.

Basic Lead-Controller: Inductive Equivalent

Just as there exists an equivalent circuit for the integral-controller that employs inductance in place of capacitance, so does there exist an inductive counterpart of the capacitive lead-controller of figure 58. The diagram of the equivalent circuit is shown in figure 84, and the transfer-function of the circuit is given by Equation (295), if source and load resistances are neglected:

$$K_{da}G_{da}(s) = k_{d2} \frac{R_1}{R_1 + R_2} \quad \frac{1 + \dfrac{L}{R_1} s}{1 + \dfrac{L}{R_1 + R_2} s} \qquad (295)$$

The form of Equation (295) is the same as that of the general expression for the basic lead-controller given by Equation (192). The two expressions are identical provided

$$k_{d2} = K_d \qquad (296)$$

$$\frac{R_1 + R_2}{R_1} = \alpha_d \qquad (297)$$

$$\frac{L}{R_1 + R_2} = \tau_d \qquad (298)$$

The limitation on the use of inductance is not as stringent in this application as in the case of the integral-controller becuase the time constant involved is shorter. Nevertheless, in most applications the capacitive circuit, figure 58, is more practical. The effect of source and load resistances in the circuit can be handled in a manner similar to that employed in previous examples.

Matching Lead-Controller

The general form for the transfer-function of the matching lead-controller is given by Equation (189).

$$K_{dc}G_{dc}(s) = k_{c1} \frac{s^2 + 2\zeta_c\omega_c s + \omega_c^2}{s^2 + 2n\zeta_c\omega_c s + \alpha_c^2\omega_c^2} \tag{249}$$

The discussion beginning with page 122 has shown that the characteristic frequency, ω_c, and the damping ratio, ζ_c, are chosen to match the characteristic frequency and damping ratio of the system that the controller is matching. The attenuation constant, α_c , and the factor, n, then determine the characteristic frequency and the damping ratio of the complete system. Thus all the parameters ζ_c, ω_c, and α_c, and n, are determined by the system with which the controller is employed and by the final servomechanism performance desired.

Figure 70 is a diagram of a circuit by means of which the function of Equation (249) may be realized. Since the constants, ζ_c, ω_c, α_c, and n are given, it is necessary to determine the values of the physical components of the circuit in terms of these constants. For the most part, this has already been done; the results are repeated here and a few others added. All of the results are readily obtained from the defining relationships listed on pages 124 and 125.

$$\frac{r_1 + r_2}{r_2} = \alpha_c^2 \tag{243}$$

$$\frac{1}{LC} = \alpha_c^2\omega_c^2 \tag{247}$$

$$\frac{R}{L} = 2n\zeta_c\omega_c \tag{248a}$$

$$m = \frac{\alpha_c^2 - n}{n(\alpha_c^2 - 1)} \tag{245a}$$

Equation (245a) or the relation (245) from which it is derived shows that the range of the factor n is from unity to α_c^2 . If the factor n has a value between unity and α_c , the damping ratio of the compensated system is reduced from that of the original system. (See Equation (256)). However, if the factor n has a value between α_c and α_c^2 the damping ratio of the complete system is increased over that of the original system. Values of n less than α_c call for large values of the adjustment factor m, while values of n greater than α_c require small values of the factor m. Decreasing the damping ratio of the original system presents no problem, since comparatively large values of m are needed; however, the minimum value of m is effectively limited by the resistance associated with the inductance, L, and thus the maximum value of n is somewhat less than α_c^2 . Therefore, the factor by which the system damping ratio can be increased is limited by the maximum indictance-to-resistance ratio that can be physically obtained in a reasonable size inductor. It is of importance to determine the maximum value of the factor n that is permitted by the ratio of L to R attainable in a reasonable size inductor.

Let R_L equal the effective resistance of the inductor and define τ_L by Equation (299).

$$\frac{L}{R_L} = \tau_L \tag{299}$$

The approximate maximum value of τ_L that can be achieved with a reasonable size inductor is one second. The minimum value of m is determined by $mR = R_L$ (300)

The maximum value of n corresponds to the above value of m and is found to be

$$n \le \frac{\alpha_c^2\,(2\zeta_c\omega_c\tau_L - 1) + 1}{2\zeta_c\omega_c\tau_L} \tag{301}$$

The inverse problem can also be solved and the minimum value of τ_L found that permits a required value of n to be secured.

$$\tau_L \ge \frac{\alpha_c^2 - 1}{2\zeta_c\omega_c(\alpha_c^2 - n)} \tag{302}$$

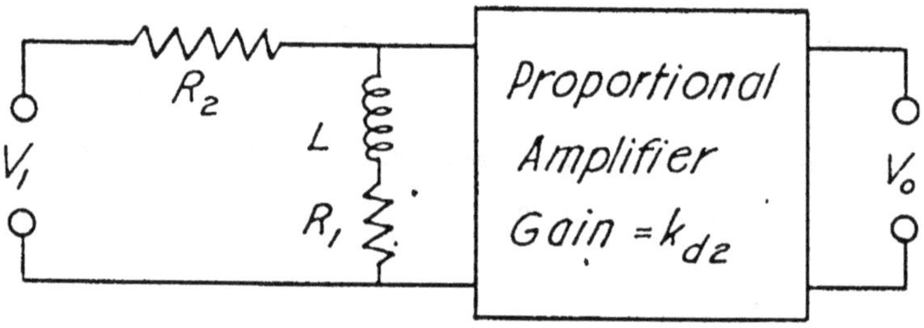

FIGURE 84

Matching Lead-Controller: Resistance of Voltage Divider Considered.

The resistances r_1 and r_2 in the matching lead-controller circuit illustrated by figure 70 were considered small enough to be neglected. If these resistances are appreciable, the characteristics of the matching lead-controller can be preserved by including the resistances in the circuit design. A circuit in which these resistances are not neglected is shown in figure 85.

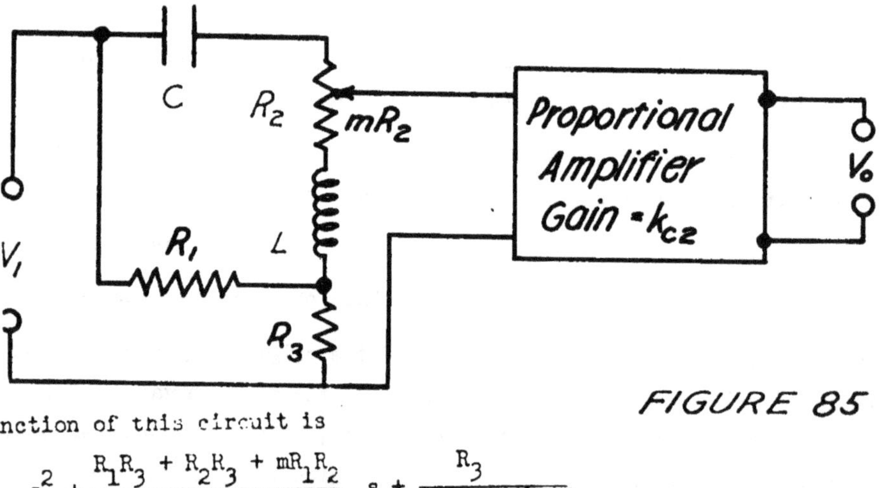

FIGURE 85

The transfer-function of this circuit is

$$K_{dc}G_{dc}(s) = \frac{s^2 + \dfrac{R_1R_3 + R_2R_3 + mR_1R_2}{(R_1 + R_3)\,L}\,s + \dfrac{R_3}{(R_1 + R_3)\,LC}}{s^2 + \dfrac{R_1R_3 + R_2R_3 + R_1R_2}{(R_1 + R_3)\,L} + \dfrac{1}{LC}} \tag{303}$$

Comparing Equation (303) with Equation (249), it is found that the two equations are equivalent provided that

$$\frac{R_1 + R_3}{R_3} = \alpha_c^2 \tag{304}$$

$$\frac{1}{LC} = \alpha_c^2 \omega_c^2 \tag{305}$$

$$\zeta_c = \frac{R_1R_3 + R_2R_3 + mR_1R_2}{2\sqrt{R_3(R_1 + R_3)}} \sqrt{\frac{C}{L}} \tag{306}$$

$$n = \frac{R_1R_3 + R_2R_3 + R_1R_2}{R_1R_3 + R_2R_3 + mR_1R_2} \tag{307}$$

Equations (304) to (307) are defining equations but are not in a form that permits facile calculation of circuit parameters from given values of α_c , ω_c, ζ_c, and n. If the equations are manipulated algebraically, the following relations are secured:

$$\frac{R_3}{R_2} = \frac{\alpha_c^2}{\alpha_c^2 - 1} \left[\frac{2n\zeta_c\omega_c L}{R_2} - 1\right] \tag{308}$$

$$\frac{R_1}{R_2} = \alpha_c^2 \left[\frac{2n\zeta_c\omega_c L}{R_2} - 1\right] \tag{309}$$

$$\frac{1}{LC} = \alpha_c^2 \omega_c^2 \tag{305}$$

$$m = 1 + \frac{\alpha_c^2}{\alpha_c^2 - 1} \left(1 - \frac{1}{n}\right) \frac{2n\zeta_c\omega_c L}{R_2} \tag{310}$$

Equations (308) and (309) show that

$$\frac{2n\zeta_c\omega_c L}{R_2} > 1 \tag{311}$$

in order that the values of R_1 and R_3 be positive and finite. On the other hand, Equation (310) demands that

$$\frac{\alpha_c^2}{\alpha_c^2 - 1} \left(1 - \frac{1}{n}\right) \frac{2n\zeta_c\omega_c L}{R_2} < 1 \tag{312}$$

in order that the value of m be positive and finite. In addition the expression on the left side of (312) must be sufficiently less than unity that the value of m is large enough that physical inductors can be employed in the circuit.

Inequalities (311) and (312) are contrasting requirements. The first demands a large value of $\frac{L}{R_2}$, while the second demands a small value of $\frac{L}{R_2}$ Actually, inequality (311) is not particularly difficult to satisfy because the left side of the relation need be only slightly larger than unity since the ratio $\frac{R_3}{R_2}$ can be very small.

Let $\frac{L}{R_2} = \frac{1 + q}{2n\zeta_c\omega_c}$ (313)

in which $q << 1$.

Then,

$$\frac{R_3}{R_2} = \frac{\alpha_c^2}{\alpha_c^2 - 1} q \tag{314}$$

$$\frac{R_1}{R_2} = \alpha_c^2 q \tag{315}$$

In addition, let R_L equal the effective resistance of the inductor, L, and let m be set at its physically minimum value of

$$mR_2 = R_L . \tag{300}$$

Again denote the ratio $\frac{L}{R_L}$ by τ_L.

A relation interconnecting q, τ_L, and the design constants ζ_c, ω_c, α_c, and n can be found and is given by Equation (316). The equation is marked approximate because simplifying approximations were made which were based upon the fact that $q<<1$. The expression can be considered correct for practical calculation as long as $q<1$.

$$n \cong \frac{\alpha_c^2 (2\zeta_c\omega_c\tau_L - 1) + 1}{2\zeta_c\omega_c\tau_L (1 + q\,\alpha_c^2)} \tag{316}$$

Since Equation (316) is based upon the minimum possible vaue of m, the value of n that it yields is the maximum value of n and thus determines the maximum factor by which the system damping ratio can be increased. The value of n approaches its theoretical maximum value of a_c^2 provided that

$$q\,\alpha_c^2 << 1 \tag{317}$$

and

$$2\zeta_c\omega_c\tau_L > 4, \text{ approximately.} \tag{318}$$

The restriction imposed by Equation (317) is easily met. The one demanded by (318) may be difficult to realize under certain circumstances.

It is reiterated that relation (318) is required only if the maximum possible value of n is sought. If the servo system is so designed that it is unnecessary to increase the damping ratio or if it is necessary to decrease the damping ratio, inequality (318) is not nearly so stringent and little difficulty is encountered in obtaining physically the required value of inductive time constant, $\mathcal{T}_L$.

Combined Basic Lead-Controller and Integral-Controller; Case I, $a_i = a_d$

When it is desired to embody both a lead-controller and an integral-controller in a servomechanism, any of the circuits described heretofore can be cascaded as long as they are isolated from one another by amplifiers of the proper design. Isolation of the circuits that have been described heretofore is generally necessary in order to prevent interaction between successive circuits that might vitiate their properties. Certain circuits, however, simultaneously perform the functions of both an under-compensating integral-controller and a basic lead-controller and in instances where both functions are desired, their use results in considerable simplification of the controller circuit.

The transfer-function of a cascaded basic lead-controller and integral-controller is found by forming the product of Equations (192) and (279). If $K_{di}G_{di}(s)$ denotes the transfer-function of the cascaded system, then

$$K_{di}G_{di}(s) = K_{da}K_iG_{da}(s)\,G_i(s) = \frac{k_dk_i}{a_d}\,\frac{(1+\mathcal{T}_da_ds)(1+\mathcal{T}_is)}{(1+\mathcal{T}_ds)(1+a_i\mathcal{T}_is)} \tag{319}$$

$$K_{di}G_{di}(s) = \frac{k_dk_i}{a_d}\,\frac{a_d\mathcal{T}_d\mathcal{T}_is^2 + (a_d\mathcal{T}_d + \mathcal{T}_i)s + 1}{a_i\mathcal{T}_d\mathcal{T}_i\,s^2 + (\mathcal{T}_d + a_i\mathcal{T}_i)\,s + 1} \tag{320}$$

The problem is to design a network whose transfer-function is equivalent to that of Equation (320). The problem is divided into three parts, depending upon the relative size of the attenuation factors a_i and a_d. The first case that is considered is the one in which a_i and a_d are equal. Accordingly, let

$$a_i = a_d = a. \tag{321}$$

Equation (320) becomes

$$K_{di}G_{di}(s) = \frac{k_dk_i}{a}\,\frac{a\,\mathcal{T}_d\mathcal{T}_is^2 + (a\,\mathcal{T}_d + \mathcal{T}_i)s + 1}{a\mathcal{T}_d\mathcal{T}_is^2 + (\mathcal{T}_d + a\mathcal{T}_i)\,s + 1}\,. \tag{322}$$

The transfer-locus of Equation (322) is a circle as illustrated by figure 87. Therefore, any network realizing the function of (322) must also be a circle.

A possible network is that illustrated by figure 88. Its transfer-locus has unit magnitude at zero frequency and infinite frequency and smaller values at intermediate frequencies. While this is no assurance that its transfer-locus is a circle, it is sufficient evidence to warrant further investigation. Accordingly, its transfer-function is determined and found to be

$$K_{di}G_{di}(s) = k_{g1}\,\frac{R_1R_2C_1C_2s^2 + (R_1C_1 + R_2C_2)s + 1}{R_1R_2C_1C_2s^2 + (R_1C_1 + R_2C_2 + R_1C_2)\,s + 1} \tag{323}$$

Equations (322) and (323) are seen to be identical provided that the following relations hold.

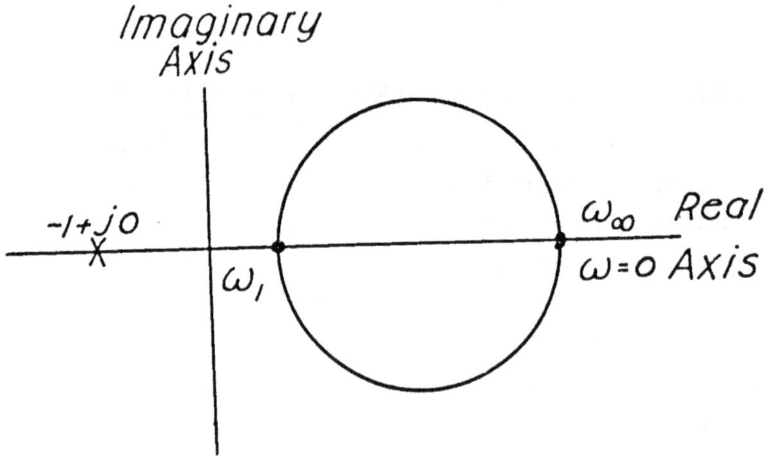

FIGURE 87

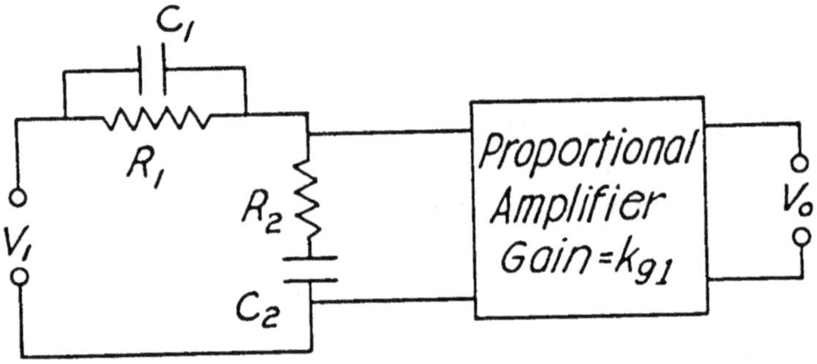

FIGURE 88

$$k_{g1} = \frac{k_d k_i}{a} \tag{324}$$

$$R_1R_2C_1C_2 = a\tau_d\tau_i \tag{325}$$

$$R_1C_1 + R_2C_2 = \alpha\tau_d + \tau_i \tag{326}$$

$$R_1C_1 + R_2C_2 + R_1C_2 = \tau_d + a\tau_i \tag{327}$$

Other than the saving in the number of components required, the chief advantage of employing the circuit of figure 88 in place of two cascaded controllers is the fact that the net amplifier gain required for the circuit of figure 83 is less by a factor, a, compared with that required by two cascaded controllers, as shown by Equation (324). This saving in gain is very important in many applications.

Equations(325), (326), and (327) may be solved provided that one circuit element is independently chosen. Let R_1 be that circuit element. When the equations are solved, it is found that two sets of values of the circuit elements, R_2, C_1, and C_2 will satisfy Equations (268) to (270). Corresponding values are not far different, however, and only one set of values is given here.

$$R_2 = \frac{\tau_i}{(\tau_i - \tau_d)(a - 1)} R_1 \tag{328}$$

$$C_1 = \frac{a\tau_d}{R_1} \tag{329}$$

$$C_2 = \frac{(\tau_i - \tau_d)(a - 1)}{R_1} \tag{330}$$

After determination of the design constants, a, τ_i, τ_d, and the system gain factor, the values of the circuit elements can be calculated at once from Equations (324), (328), (329), and (330).

An inductive equivalent circuit exists for the circuit of figure 88, and is illustrated by figure 89. The values of the circuit elements are easily found from given values of the design parameters, k_1, a, τ_d, and τ_i

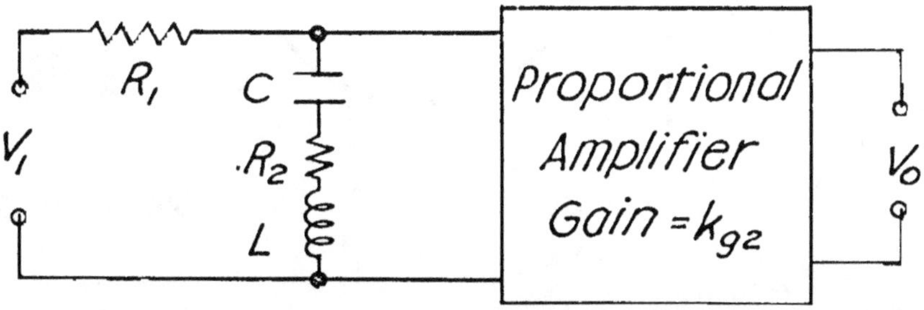

FIGURE 89

Combined Basic Lead-Controller and Integral-Controller; Case II, $a_i > a_d$

The circuit of figure 88 is incapable of simulating a cascaded lead-controller and integral-controller if the attenuation constants of the lead-controller and integral-controller are unequal. The circuit can be so modified, however, that such a condition can be realized. Since the circuit modification depends upon whether a_i or a_d is the larger, the two cases are handled individually. The first case considered is the one in which the attenuation factor of the integral-controller is larger than that of the lead-controller.

The transfer-locus of a cascaded controller in which $a_i > a_d$ is illustrated in figure 90, and the modified circuit that possesses this form of transfer-locus is illustrated in figure 91. The modification is the addition of resistance in series with C_1, the capacitance that is responsible for the introduction of phase-lead, in the transfer-function. The transfer-function of the modified circuit is given by Equation (331).

$$K_{di}G_{di}(s) = k_{g2} \frac{R_2C_2(R_1 + R_c)\, C_1 s^2 + \left[R_2C_2 + (R_1 + R_c)\, C_1\right] s + 1}{\left[R_2C_2(R_1 + R_c)\, C_1 + R_1R_cC_1C_2\right] s^2 + \left[R_2C_2 + (R_1 + R_c)\, C_1 + R_1C_2\right] s + 1} \tag{331}$$

Equations (320) and (331) are similar in form and are equivalent provided that the following relations hold:

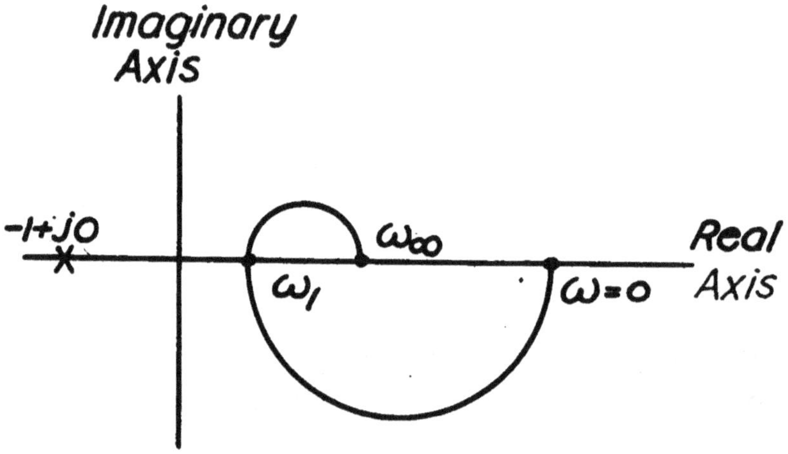

FIGURE 90

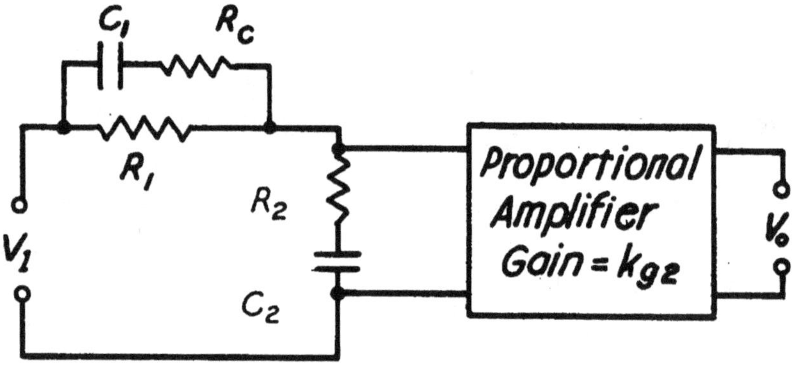

FIGURE 91

$$\frac{k_dk_i}{a_d} = k_{g2} \tag{332}$$

$$a_d\tau_d\tau_i = R_2C_2\,(R_1 + R_c)\,C_1 \tag{333}$$

$$a_i\tau_d\tau_i = R_2C_2(R_1 + R_c)\,C_1 + R_1R_cC_1C_2 \tag{334}$$

$$\tau_i + a_d\tau_d = R_2C_2 + (R_1 + R_c)\, C_1 \tag{335}$$

$$a_i\tau_i + \tau_d = R_2C_2 + (R_1 + R_c)\, C_1 + R_1C_2 \tag{336}$$

The constants a_d, τ_d, τ_i, and a_i are determined by methods indicated heretofore. After their values have been chosen, the circuit of figure 91 may be designed. It is desirable, therefore, that the value of each circuit element by expressed in terms of the known quantities, a_i, a_d, τ_i, and τ_d. These expressions are found by solving Equations (333) to (336) simultaneously. However, since there are five unknown circuit elements and four determining equations, one circuit element must be chosen from other considerations and the values of the other circuit components determined in terms of the value of the separately chosen element. Any one of the circuit elements may be so chosen but a particularly convenient choice is the resistance R_1. Its value can be so selected that the remainder of the circuit elements have reasonable values. The solution of Equations (333) to (336) is given below:

$$R_c = \frac{\tau_i(a_i - a_d)\, R_1}{(a_d - 1)\,(\tau_i a_i - \tau_d a_d)} \tag{337}$$

$$R_2 = \frac{\tau_i R_i}{\tau_i\,(a_i - 1) - \tau_d\,(a_d - 1)} \tag{338}$$

$$C_2 = \frac{\tau_i(a_i - 1) - \tau_d(a_d - 1)}{R_1} \tag{339}$$

$$C_1 = \frac{(a_d - 1)(\tau_i a_i - \tau_d a_d)\tau_d}{\left[\tau_i(a_i - 1) - \tau_d\,(a_d - 1)\right] R_1} \tag{340}$$

Equation (332) and Equations (337) to (340) completely specify the controller of figure 91. The relations (337) to (340) reduce to the corresponding equations (328) to (330) if a_i is set equal to a_d and their common value denoted by a.

<u>Combined Basic Lead-Controller and Integral-Controller:</u> <u>Case III, $a_d > a_i$</u>

The circuit of figure 91 is unsatisfactory if the system design demands that the attenuation factor of the lead-controller be larger than that of the integral-controller. A circuit of the type illustrated by figure 92, however, will realize the required transfer-function. The transfer-function of this circuit is given by Equation (341) and the transfer-locus is illustrated by figure 93.

$$K_{di}G_{di}(s) = \frac{(R_2+R_p)k_{g3}}{R_1+R_2+R_p}\;\frac{\dfrac{R_1R_2R_pC_1C_2}{R_2+R_p}\,s^2 + \dfrac{R_1R_2C_1+R_2R_pC_2+R_1R_pC_1}{R_2+R_p}\,s + 1}{\dfrac{R_1R_2R_pC_1C_2}{R_2+R_p}\,s^2 + \dfrac{R_1R_2C_1+R_2R_pC_2+R_1R_pC_1+R_1R_pC_2}{R_1+R_2+R_p}\,s + 1} \tag{341}$$

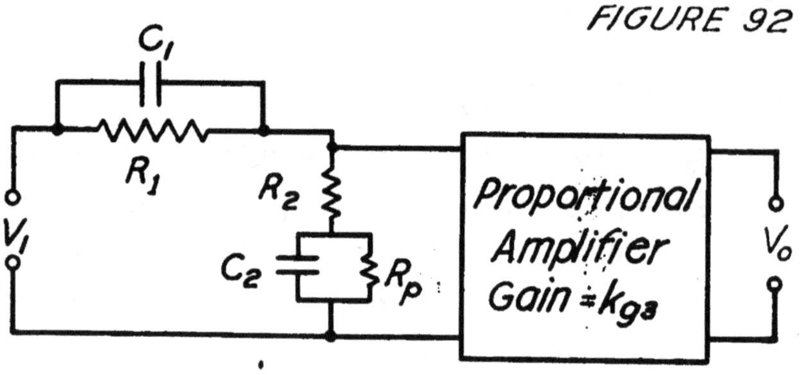

FIGURE 92

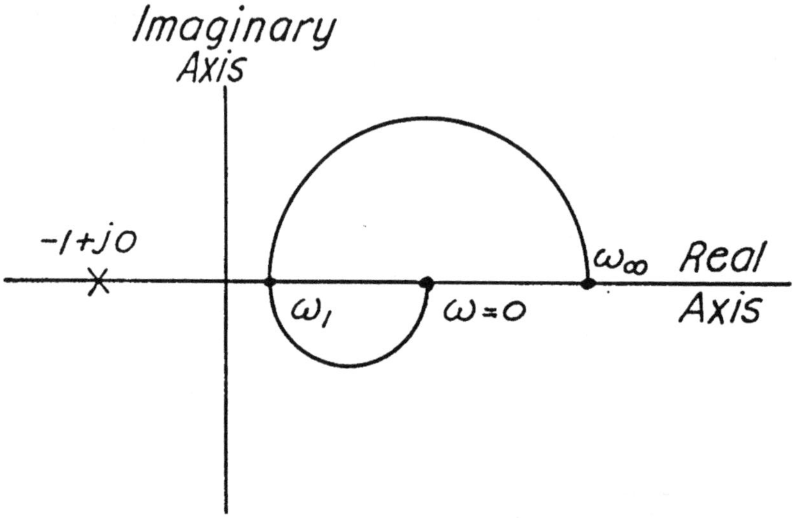

FIGURE 93

Equations (341) and (320) are equivalent provided that

$$\frac{k_d k_i}{\alpha_d} = \frac{(R_2 + R_p)\, k_{g3}}{R_1 + R_2 + R_p} \tag{342}$$

$$\alpha_d \tau_d \tau_i = \frac{R_1 R_2 R_p C_1 C_2}{R_2 + R_p} \tag{343}$$

$$\alpha_i \tau_d \tau_i = \frac{R_1 R_2 R_p C_1 C_2}{R_1 + R_2 + R_p} \tag{344}$$

$$+ \alpha_d \tau_d = \frac{R_1 R_2 C_1 + R_2 R_p C_2 + R_1 R_p C_1}{R_2 + R_p} \tag{345}$$

$$a_i\tau_i + \tau_d = \frac{R_1R_2C_1 + R_2R_pC_2 + R_1R_pC_1 + R_1R_pC_2}{R_1 + R_2 + R_p} \qquad (346)$$

Equations (343) to (346) can be solved simultaneously and the values of the circuit elements determined in terms of the design parameters. The solutions are given below.

$$R_p = \frac{a_ia_d\,(a_i - 1)(\tau_i - \tau_d)\,R_1}{(a_d - a_i)\left[a_i\tau_i(a_d - 1) - a_d\tau_d(a_i - 1)\right]} \qquad (347)$$

$$R_2 = \frac{a_i\tau_iR_1}{a_i\tau_i\,(a_d - 1) - a_d\tau_d\,(a_i - 1)} \qquad (348)$$

$$C_1 = \frac{a_d\,\tau_d}{R_1} \qquad (349)$$

$$C_2 = \frac{\left[a_i\tau_i(a_d - 1) - a_d\tau_d\,(a_i - 1)\right]^2}{a_ia_d\,(a_i - 1)(\tau_i - \tau_d)\,R_1} \qquad (350)$$

After the values of R_p and R_2 are determined from Equations (347) and (348) the proper gain, k_{g3}, of the amplifier associated with the circuit may be calculated from Equation (342a).

$$k_{g3} = \frac{(R_1 + R_2 + R_p)}{R_2 + R_p}\;\frac{k_dk_i}{a_d} \qquad (342a)$$

CHAPTER X

REGENERATIVE CONTROLLERS

Chapter V, devoted to the theory of minimum velocity-error systems, has shown that since an ideal integral-controller possesses infinite gain at zero frequency, it can be realized physically only through the use of regenerative circuits. A stable circuit for the ideal integral-controller was developed, and its proper adjustment was discussed. It was also shown that this circuit could be so adjusted that it was no longer an ideal integral-controller, but became either an over-compensating integral-controller or an under-compensating integral-controller, depending upon whether the loop gain was greater or less than that required to give infinite gain at zero-frequency to the controller. It was also shown that the under-compensating integral-controller could be realized by a network cascaded with an amplifying element. The remainder of the circuits that have been discussed in this paper have also been circuits in which the amplifying element was cascaded with the network controlling the characteristics of the device.

Although regenerative circuits are required for the physical realization of ideal integral-controllers only, they are frequently useful in the design of lead-controllers and of combined lead- and integral-controllers. In such applications, the regeneration factor is so adjusted that the gain is high but not so high that ageing of the circuit elements might cause the controller to become unstable. Such use of regeneration to obtain a high gain controller frequently reduces the number of stages of amplification necessary in the controller and thus simplifies the servo system.

This chapter describes regenerative circuits for realizing several control functions. Detailed description of all circuits of this type is unnecessary, since once the general design procedure is established, the manner by which variations in the controller characteristics are secured becomes evident.

Under-Compensating Integral-Controller

The design of a regenerative circuit that will behave as an under-compensating integral-controller has been explained in Chapter V. The results are recapitulated here as an introduction to the basic form of the regenerative circuit which is employed in this chapter to realize other controller characteristics.

Figure 52, reproduced below for convenience, is the block diagram of the basic circuit. L(s) denotes the transfer-function of the feedback path containing the network which determines the controller characteristic. The transfer-function of the complete controller is represented by $K_rG_r(s)$, in which the subscript "r" denotes a regenerative controller. A second subscript following the r distinguishes the various characteristic functions secured by the application of the regenerative circuit.

The general relation interconnecting the transfer-function of the complete controller, the transfer-function of the feedback path, and the regeneration factor, h, is

$$K_rG_r(s) = \frac{k_p}{1 - hL(s)} \tag{160a}$$

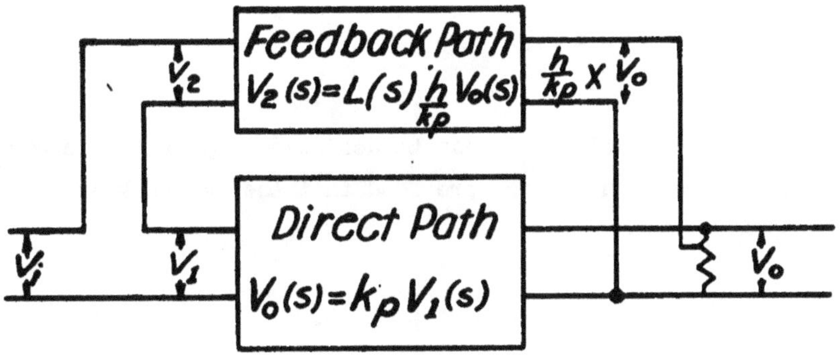

FIGURE 52

It has been shown that if the feedback path contains an RC network such as is illustrated in figure 94, its transfer-function, L(s), is

$$L(s) = \frac{1}{1 + RCs} \tag{137}$$

and the transfer-function of the controller becomes

$$K_{ri}G_{ri}(s) = \frac{k_p}{1 - h} \quad \frac{1 + RCs}{1 + \dfrac{RC}{1 - h} s} \tag{351}$$

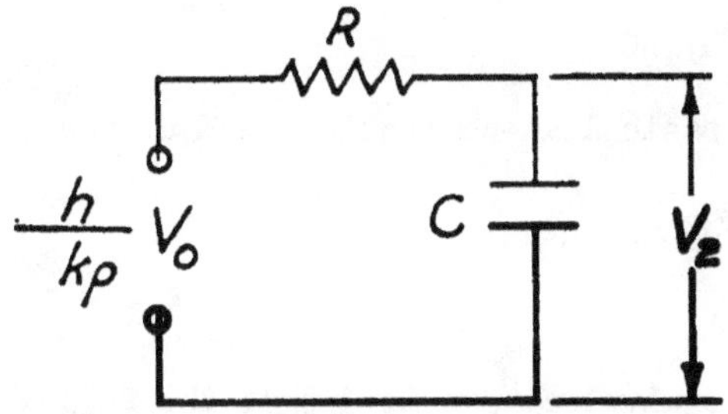

FIGURE 94

The prototype function of the under-compensating integral-controller is given by Equation (279):

$$K_i G_i(s) = k_i \quad \frac{1 + \tau_i s}{1 + a_i \tau_i s} \tag{279}$$

It is evident that Equations (279) and (351) are identical provided that,

$$k_i = \frac{k_p}{1 - h} \tag{352}$$

$$\tau_d = \frac{RC}{1 - h} \tag{353}$$

$$\alpha_d = \frac{1}{1 - h} \qquad (354)$$

Basic Lead-Controller

The regenerative circuit of figure 52 can be designed to possess a characteristic equivalent to that of the basic lead-controller, provided that the feedback network is of the form illustrated by figure 95.

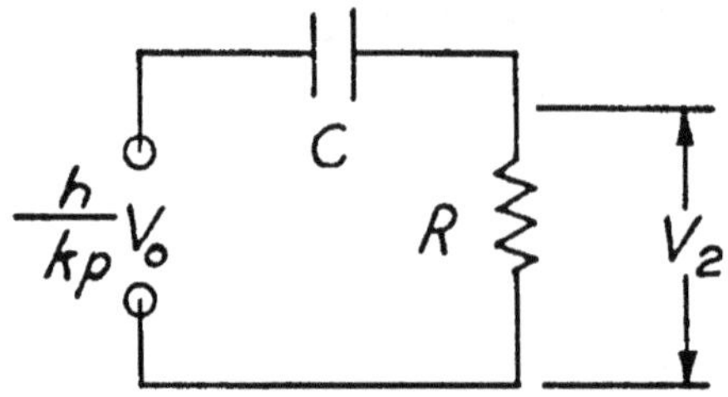

FIGURE 95

The transfer-function, L(s), of this circuit is

$$L(s) = \frac{RCs}{1 + RCs} \qquad (355)$$

which upon substitution into Equation (160a) yields Equation (356) as the overall transfer-function of the controller.

$$K_{rd}G_{rd}(s) = k_p \frac{1 + RCs}{1 + (1 - h)\, RCs} \qquad (356)$$

The prototype function of the basic lead-controller is Equation (192), page 96.

$$K_{da}G_{da}(s) = \frac{k_d}{\alpha_d} \frac{1 + \tau_d \alpha_d s}{1 + \tau_d s} \qquad (192)$$

Equations (192) and (356) are equivalent provided that the following relations are maintained:

$$k_p = \frac{k_d}{\alpha_d} \qquad (357)$$

$$(1 - h)\, RC = \tau_d \qquad (358)$$

$$\frac{1}{1 - h} = \alpha_d \qquad (359)$$

Matching Lead-Controller

The regenerative controller will realize the characteristics of a matching lead-controller, provided that the feedback network has the form illustrated by figure 96. The transfer-function of this network is given by Equation (360).

$$L(s) = \frac{LCs^2 + mRCs}{LCs^2 + RCs + 1} \qquad (360)$$

The transfer-function of the controller, found by substituting Equation (360) into Equation (160a), is given by Equation (361)

$$K_{rc}G_{rc}(s) = \frac{k_p}{1 - h} \; \frac{s^2 + \frac{R}{L} s + \frac{1}{LC}}{s^2 + \frac{R(1 - hm)}{L(1 - h)} s + \frac{1}{(1 - h)\, LC}} \tag{361}$$

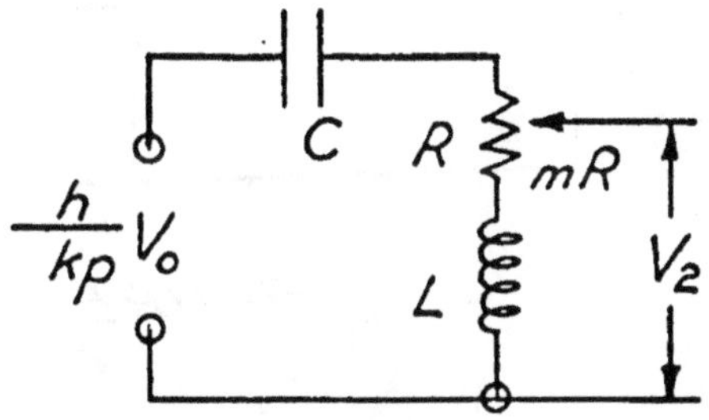

FIGURE 96

The prototype function of the matching phase-lead controller is given by Equation (249).

$$K_{dc}G_{dc}(s) = k_{c1} \frac{s^2 + 2\zeta_c\omega_c s + \omega_c^2}{s^2 + 2\zeta_c n\omega_c^2 + \alpha_c^2\omega_c^2} \tag{249}$$

Equations (249) and (361) are equivalent provided that the following relations hold:

$$\frac{k_p}{1 - h} = k_{c1} \tag{362}$$

$$\frac{1}{LC} = \omega_c^2 \tag{363}$$

$$\frac{1}{1 - h} = \alpha_c^2 \tag{364}$$

$$\frac{R}{L} = 2\zeta_c\omega_c \tag{365}$$

$$\frac{1-hm}{1-h} = n \tag{366}$$

Combined Lead-Controller and Integral-Controller; Attenuation Factors Equal

The feedback link of the regenerative controller can be so designed that it simulates the response of combined lead- and integral-controllers. The necessary form of feedback network is illustrated by figure 97.

This network is the same as that employed in the cascade circuit for achieving a combined lead- and integral-controller (See figure 88). The transfer-function of the network is:

$$L(s) = \frac{R_1R_2C_1C_2s^2 + (R_1C_1 + R_2C_2)\, s + 1}{R_1R_2C_1C_2s^2 + (R_1C_1 + R_2C_2 + R_1C_2)\, s + 1} \tag{367}$$

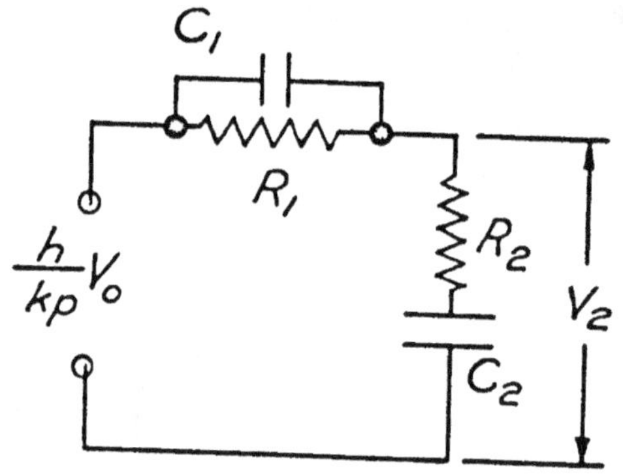

FIGURE 97

Equation (367) can be written

$$L(s) = \frac{As^2 + Bs + 1}{As^2 + Ds + 1} \tag{368}$$

The transfer-function of the complete controller is obtained by substituting Equation (368) into Equation (160a):

$$K_{rdi}G_{rdi}(s) = \frac{k_p}{1 - h} \quad \frac{As^2 + Ds + 1}{As^2 + \frac{(D - Bh)}{1 - h} s + 1} \tag{372}$$

The prototype function for the combined integral- and lead-controller with equal attenuation factors is Equation (322).

$$K_{di}G_{di}(s) = \frac{k_d k_i}{\alpha} \quad \frac{\alpha\tau_d\tau_i s^2 + (\alpha\tau_d + \tau_i) s + 1}{\alpha\tau_d\tau_i s^2 + (\tau_d + \alpha\tau_i) s + 1} \tag{322}$$

Equations (322) and (372) are equivalent provided that

$$\frac{k_p}{1 - h} = \frac{k_d k_i}{\alpha} \tag{373}$$

$$A = \alpha\tau_d\tau_i \tag{374}$$

$$D = \alpha\tau_d + \tau_i \tag{375}$$

$$\frac{D - Bh}{1 - h} = \tau_d + \alpha\tau_i \tag{376}$$

The above equations can be solved simultaneously and the values of the circuit elements found.

The above regenerative controller simulates a cascaded integral- and lead-controller in which the attenuation constants are equal. The circuit of the feedback link can be modified in the fashion indicated in the preceding chapter to physically realize a control function with unequal attenuation factors.

CHAPTER XI

CONSTANT-RESISTANCE CONTROLLERS

It is frequently necessary to synthesize a controller whose transfer-function is the product of several of the compensating functions defined by Equations (192), (249), and (279). A solution to this problem consists of physically realizing each compensating function with one of the several possible networks discussed and then so combining these individual networks that no interaction occurs between them. The circuits thus far considered require some type of buffer amplifier between networks to ensure that the transfer-function of the combined system is actually the product of the transfer-functions of the individual sections. Thus a basic lead-controller and an integral-controller generally must be combined as indicated by figure 98b and not as shown by figure 98a. A buffer amplifier is unnecessary provided that the impedance looking toward the output terminals from the inter-connection point is several times the impedance looking toward the input terminals from that same point. Thus, the circuit of figure 98a may be connected as illustrated provided that when the connections are broken at XX, the impedance Z_0 is several times the impedance Z_1. It is often difficult to meet this condition, however, and in those instances it is better to employ a buffer amplifier when interconnecting any of the circuits discussed heretofore.

A type of network exists, however, which can be cascaded with another network of the same type without encountering interaction between networks. A network of this type is termed a "constant resistance" network because the impedance seen looking into the network from either the output or input terminals is resistive and independent of frequency when the opposite end is

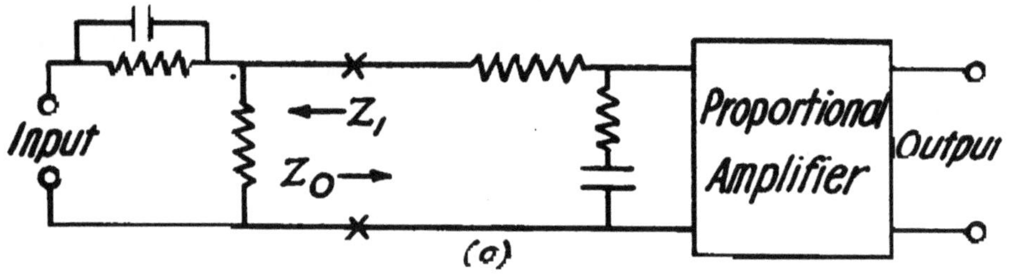

FIGURE 98a

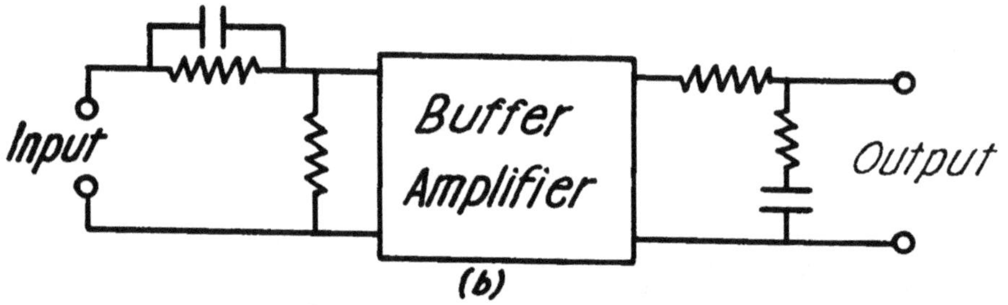

FIGURE 98b

properly terminated. For example, if the network within the square of figure 99 is a constant-resistance network, the impedance Z_1, looking into the network from input terminals aa, is equal

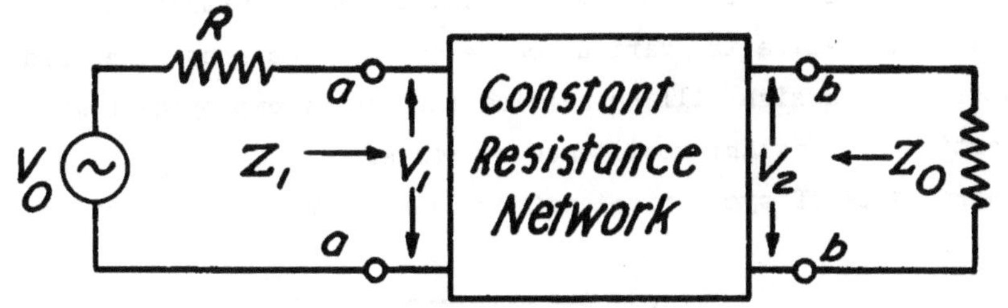

FIGURE 99

to R and the impedance Z_0 looking back into the network from terminals bb is also equal to R. The characteristics of these networks are described fully in texts on filter theory. See in particular Communication Networks, Vol. 1 and 2, by E. A. Guillemin.[10]

Certain restrictions upon the values of the elements of constant resistance networks are relevant to the design of these networks. The terminating impedance and the source impedance must both be resistances and equal to the characteristic resistance of the network. If R denotes the common value of the characteristic resistance and the terminating resistances of the network of figure 99, the impedances Z_1 and Z_0 are also both equal to R. It is obvious that the terminating resistance R may be removed and the terminals of a second constant-resistance network whose characteristic resistance and whose terminating resistance are also equal to R may be connected without disturbing the transfer-function, $\frac{V_2}{V_1}(s)$, of the original network. Such a connection is illustrated by figure 100.

FIGURE 100

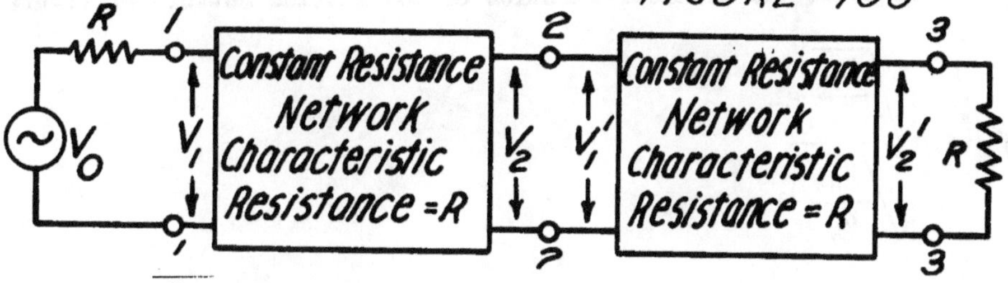

If the transfer-function $\frac{V_2}{V_1}(s)$ of the first network alone is $K_1G_1(s)$ and the transfer-function $\frac{V_2'}{V_1'}(s)$ of the second network is $K_2G_2(s)$, the transfer-function of the combined system is $K_1K_2G_1(s)\ G_2(s)$.

This property of constant-resistance networks is most useful. The networks can be designed individually and cascaded in any desired fashion with the assurance that the transfer-function of the complete system will be the product of the transfer-functions of the individual sections, provided that the characteristic resistances of the individual networks are equal and the complete system is properly terminated. Constant-resistance networks have immediate application in the synthesis of servo-controllers whose transfer-function is the product of several compensating functions. It is the purpose of this chapter to illustrate how constant-resistance networks are found that realize the various compensating functions discussed in the earlier portions of this paper. The design will be carried through by employing symmetrical networks to realize the properties of constant-resistance networks.

The most general type of symmetrical network has a symmetrical lattice structure as illustrated in figure 101.

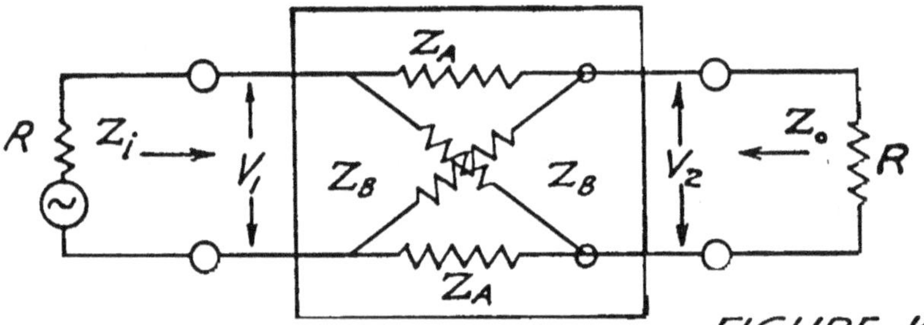

FIGURE 101

It can be readily shown that such a network has constant-resistance properties provided that

$$\sqrt{Z_A Z_B} = R \tag{377}$$

If relation (377) is fulfilled, the impedances Z_i and Z_0, seen by looking into the network from the input and output terminals, are

$$Z_i = R \tag{378}$$

$$Z_0 = R \tag{379}$$

If Equation (377) is true, the transfer-function of the lattice network of figure 101 is:

$$\frac{V_2}{V_1}(s) = \frac{\sqrt{\frac{Z_B}{Z_A}} - 1}{\sqrt{\frac{Z_B}{Z_A}} + 1} \tag{380}$$

Equations (377) and (380) are the defining relations of the lattice network. The product of the two impedances Z_B and Z_A is maintained equal to a constant whose value is equal to the terminating resistances of the network, while the ratio of Z_B/Z_A is adjusted to secure the desired transfer-function for the network. Although the lattice network itself does not lend itself readily to incorporation in practical amplifiers because it is a "balanced" network with respect to ground, the simplicity of the defining relations (377) and (380) make it useful as a design medium. After a lattice structure has been designed that realizes the desired compensating function, the lattice itself can be frequently transformed into an unbalanced structure that lends itself more readily to incorporation into a practical controller.

The development of the constant-resistance lattice network is aided if the lattice elements are "non-dimensionalized" with respect to the characteristic resistance of the network. This is done by means of the following transformations:

$$z_a = \frac{Z_A}{R} \tag{381}$$

$$z_b = \frac{Z_B}{R} \tag{382}$$

The defining equations of the lattice network, Equations (377) and (380), become:

$$z_a z_b = 1 \tag{383}$$

$$\frac{V_2}{V_1}(s) = \frac{\sqrt{\dfrac{z_b}{z_a}}}{\sqrt{\dfrac{z_b}{z_a}} + 1} \tag{384}$$

Since

$$z_b = \frac{1}{z_a} \tag{383a}$$

Equation (384) reduces to

$$\frac{V_2}{V_1}(s) = \frac{1 - z_a}{1 + z_a} \tag{385}$$

The chapter "Simulative and Corrective Networks", vol. 2, <u>Communication Networks</u> by Guillemin develops design methods that are directly applicable to the problem of synthesizing servomechanism compensating functions. It is there shown that the locus of $\frac{V_2}{V_1}(j\omega)$ is a circle provided the locus of the impedance z_a is a circle. Since several compensating functions of importance in servomechanism design have circular loci or are formed of functions with circular loci, it is important to determine the nature of z_a that results in (a), a circular impedance locus of z_a and (b) the desired type of circular locus for the network.

The above reference shows that the locus of z_a is a circle provided that z_a has the form illustrated by figure 102a in which X is a non-dissipative reactance, and r_0 and r are pure resistances. The expression for z_a is Equation (386).

$$z_a = \frac{r_0 r + X(r_0 + r)}{r + X} \tag{386}$$

Figure 102b illustrates the form of z_b which is the inverse impedance of z_a. The transfer function is obtained in terms of the components r_0, r, and X by substituting (386) into (385).

$$\frac{V_2}{V_1}(s) = -\frac{r(r_0 - 1) + X(r_0 + r - 1)}{r(r_0 + 1) + X(r_0 + r + 1)} \qquad (387)$$

The negative sign preceding the expression on the right side of Equation (387) can be disregarded since it can be cancelled by a similar phase reversal at any other point in the servo controller. Equation (387) may be written

$$\frac{V_2}{V_1}(s) = \frac{(r_0 - 1)}{r_0 + 1} \; \frac{1 + \dfrac{r_0 + r - 1}{r(r_0 - 1)} X}{1 + \dfrac{r_0 + r + 1}{r(r_0 + 1)} X} \qquad (387a)$$

In order to establish the equivalence of constant-resistance networks and other types of compensating controllers previously considered, it is necessary to cascade a proportional amplifier of adjustable gain with the lattice network of figure 101. Such a controller, illus-

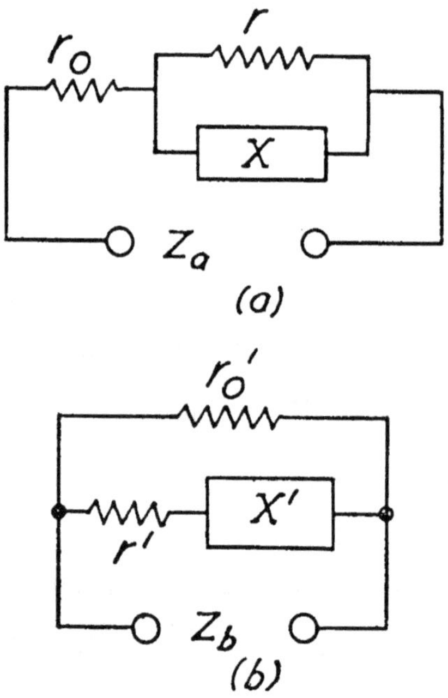

FIGURE 102

trated by figure 103, permits the gain-factor of the controller to be set at that value necessary to secure the equivalence of the constant-resistance controller with the other types of controllers described earlier. The transfer-function $K_L G_L(s)$ of this system is

$$K_L G_L(s) = \frac{V_0}{V_1} = k_L \frac{1 - z_a}{1 + z_a} \qquad (388)$$

FIGURE 103

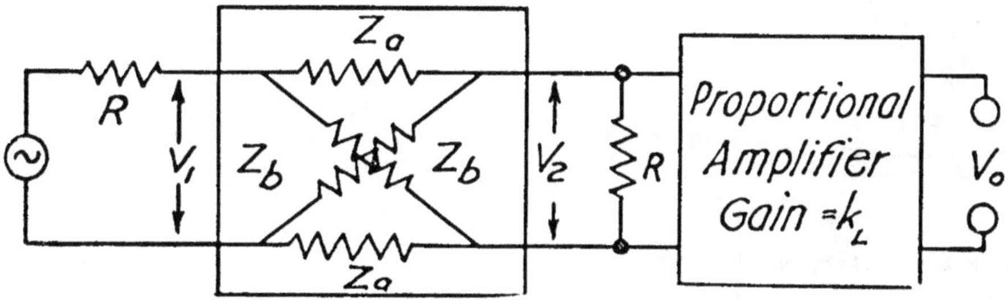

If z_a is a network whose form is illustrated by figure 102a, Equation (388) becomes

$$K_L G_L(s) = k_L \frac{(r_0 - 1) + \dfrac{r_0 + r - 1}{r} X}{(r_0 + 1) + \dfrac{r_0 + r + 1}{r} X} \qquad (389)$$

The problem is to so choose the reactance X that Equation (389) is equivalent to the compensating function that it is desired to physically realize. The nature of the reactance X is determined for several types of compensating functions in the following sections of this chapter.

Under-Compensating Integral-Controller: Constant-Resistance Network

The general form of the compensating function employed to secure under-compensating integral-control is given by Equation (279).

$$K_i G_i(s) = k_i \frac{1 + \tau_i s}{1 + a_i \tau_i s} \qquad (279)$$

Equation (279) may be realized by the constant-resistance lattice provided that X, the reactance in figure 102, is capacitive. The required form of z_a is illustrated by figure 104.

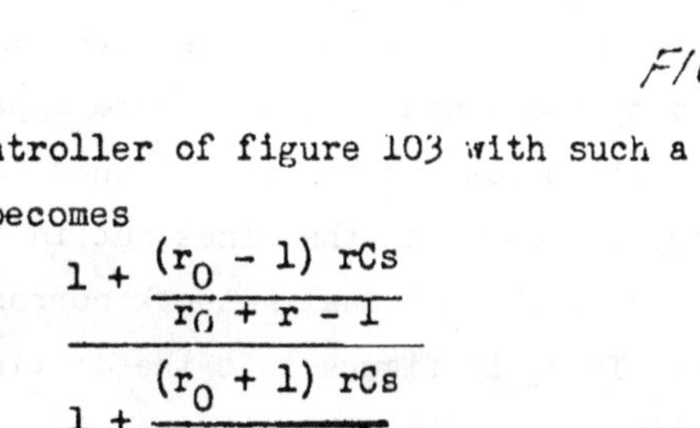

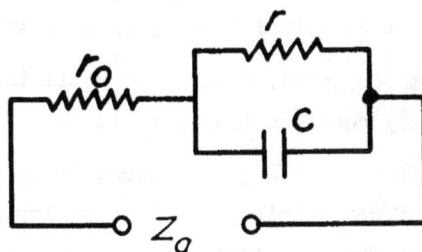

FIGURE 104

The transfer-function of the controller of figure 103 with such a network for z and its i verse (described below) as z_b, becomes

$$K_{Li} G_{Li}(s) = k_{Li} \frac{r_0 + r - 1}{r_0 + r + 1} \; \frac{1 + \dfrac{(r_0 - 1)\, rCs}{r_0 + r - 1}}{1 + \dfrac{(r_0 + 1)\, rCs}{r_0 + r + 1}} \qquad (390)$$

Equations (279) and (390) are identical provided

$$k_i = k_{L1} \frac{r_0 + r - 1}{r_0 + r + 1}, \tag{391}$$

$$\tau_i = \frac{(r_0 - 1}{r + r_0 - 1} rC \tag{392}$$

$$a_i = \frac{(r + r_0 - 1)(r_0 + 1)}{(r + r_0 + 1)(r_0 - 1)} . \tag{393}$$

The resistance r_0 must be larger than, but approximately equal to, unity. The resistance r should be several times the resistance of r_0 .

$$r \gg r_0 \tag{394}$$

$$\left.\begin{matrix} r_0 > 1 \\ r_0 \cong 1 \end{matrix}\right\} \tag{395}$$

If these conditions written in inequality form in relations (394) and (395) are maintained, the following relations are approximately correct.

$$k_i \cong k_{L1} \tag{391a}$$

$$\tau_i \cong (r_0 - 1)C \tag{392a}$$

$$a_i \cong \frac{r_0 + 1}{r_0 - 1} \tag{393a}$$

The values of k_i, a_i, and τ_i are selected in order to produce optimum servo response as described in earlier chapters. After their values are known, the circuit elements of the impedance, z_a, can be calculated from Equations (391a), (392a), and (393a).

The lattice network is not completely determined until the form and component values of the impedance, z_b, are known. By Equation (383), z_b is the inverse of the impedance z_a. The process of determining the reciprocal of a given impedance is described in most texts on circuit theory. A simple method of determining this inverse impedance is illustrated by figure 105. If figure 105a is the network whose inverse is to be found, one point is established within each loop of the network and one external to the network, and lines drawn connecting the several points. These lines comprise a map of the inverse impedance. The inverse impedance is formed by connecting along the lines between the established points impedances whose values are equal to the inverse of the impedances which the lines cut in the original network. The line cutting the driving point source of the original network corresponds to the driving point source of the inverse network. Thus, in figure 105b the driving points of the inverse network are points a, b, because the line connecting these points in figure 105a cuts the driving-point source. Following the method just outlined, a resistance connects the points ab in the inverse network whose value is the inverse of the resistance, r_0, cut by that line in the original network. Similarly, a series circuit also connects the points ab in the inverse network, made up of (1) an inductance between a and c numerically equal to the capacitance C cut by the corresponding line a,c, and (2) a resistance between points c and b equal to the inverse of the resistance cut by the corresponding line in the original network, figure 105a.

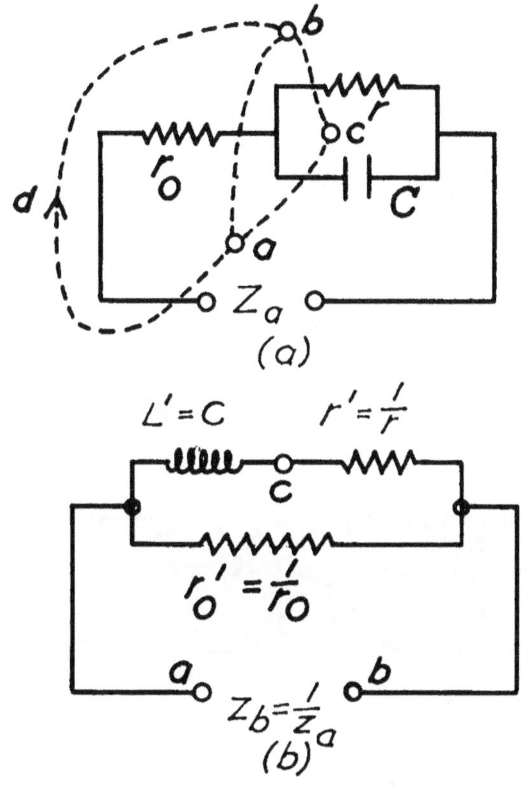

FIGURE 105

Figure 105a is an illustration of one of the lattice impedances necessary to realize an integral-controller, and the network of figure 105b is its inverse. The problem, therefore, of designing a constant-resistance integral-controller is completely resolved and the results are summarized below.

The components of z_a, figure 104, are given by,

$$\frac{r_0 - 1}{r + r_0 - 1} \, rC = \tau_1 \tag{392}$$

$$\frac{(r + r_0 - 1)(r_0 + 1)}{(r + r_0 + 1)(r_0 - 1)} = \alpha_1 \tag{393}$$

The components of z_b, figure 105b, are specified by,

$$r' = \frac{1}{r}, \tag{396}$$

$$L' = C, \tag{397}$$

$$r_0' = \frac{1}{r_0} \tag{398}$$

in which r, r_0, and C, are determined from Equations (392) and (393). The gain of the proportional amplifier cascaded with the network is

$$k_{L1} = \frac{r_0 + r + 1}{r_0 + r - 1} k_i \tag{391a}$$

Basic Lead-Controller; Constant-Resistance Network

A basic lead-controller may be realized in the form of a constant-resistance network by employing an inductance as the reactive element in the impedance z_a, as illustrated in figure 106. The expression for the transfer-function of the resulting controller is given by Equation (399).

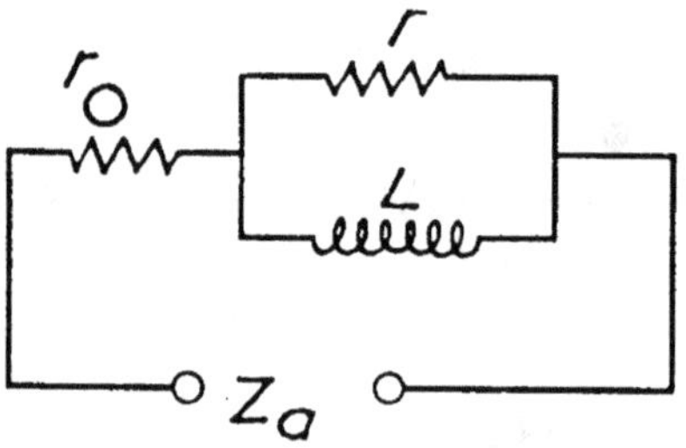

FIGURE 106

$$K_{Ld}G_{Ld}(s) = k_{L2}\,\frac{r_0 - 1}{r_0 + 1}\;\frac{1 + \dfrac{r_0 + r - 1}{r(r_0 - 1)}Ls}{1 + \dfrac{r_0 + r + 1}{r(r_0 + 1)}Ls} \tag{399}$$

The prototype equation for the basic lead-controller is given by Equation (192).

$$K_{da}G_{da}(s) = \frac{k_d}{\alpha_d}\;\frac{1 + \tau_d\alpha_d s}{1 + \tau_d s} \tag{192}$$

Equations (192) and (399) are equivalent provided that

$$k_{L2}\,\frac{r_0 - 1}{r_0 + 1} = \frac{k_d}{\alpha_d} \tag{400}$$

$$\frac{r_0 + r + 1}{r(r_0 + 1)} L = \tau_d \tag{401}$$

$$\frac{(r_0 + r - 1)(r_0 + 1)}{(r_0 + r + 1)(r_0 - 1)} = \alpha_d \qquad (402)$$

The inverse impedance, z_b, is found by the method previously described. Its form is illustrated by figure 107, and the value of the network components are given by Equations (403), (404) and (405).

$$r'_0 = \frac{1}{r_0} \qquad (403)$$

$$r' = \frac{1}{r} \qquad (404)$$

$$C' = L \qquad (405)$$

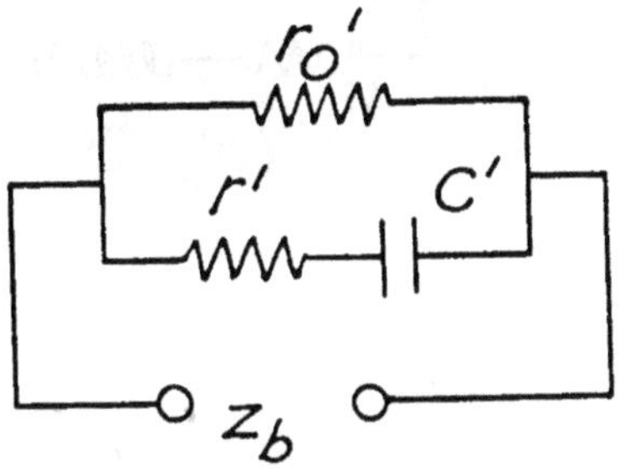

FIGURE 107

The network of figure 106 is not an entirely satisfactory form for the impedance z_a because it employs a non-resistive inductor. Fortunately, a circuit that employs a dissipative inductor exists and is exactly equivalent to the circuit of figure 106. This network is illustrated by figure 108 in which figure 108a is the diagram of the original network and figure 108b is the diagram of the equivalent circuit. The relations that must be maintained are given by Equations (406, (407), and (408).

$$r_0 = \frac{R_0R + R_0R_d + RR_d}{R + R_d} \qquad (406)$$

$$r = \frac{R^2}{R + R_d} \qquad (407)$$

$$L = L_d \frac{R^2}{(R + R_d)^2} \qquad (408$$

The above three relations show that it is possible to physically realize the single-stage lead function, Equation (136), with a physical, dissipative inductor as well as with a non-dissipitive inductance, provided the resistance R_d is not too large.

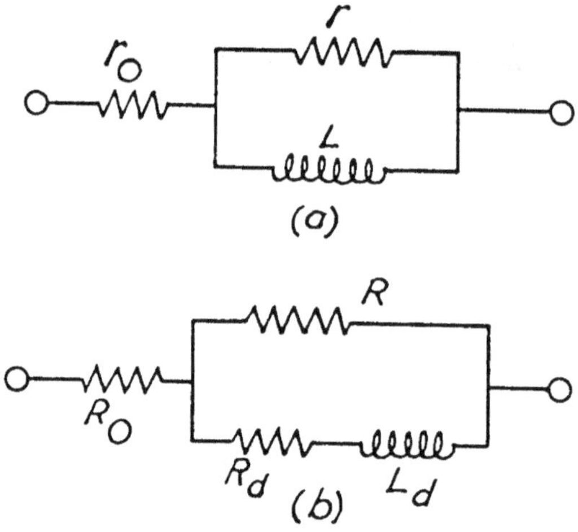

FIGURE 108

Combined Integral- and Basic Lead-Controller; Constant-Resistance Network

Constant-resistance network controllers can be designed that will simulate a combined integral-and lead-controller by (a) cascading two lattice sections if the attenuation factors α_i and α_d are unequal, or (b) realizing both compensating functions in a single lattice structure if the attenuation constants α_i and α_d are equal. The case of equal attenuation factors can also be physically realized by cascading two lattice networks, but there is seldom anything gained by such a procedure and a great deal is lost from the standpoint of controller simplicity.

The form of z_a necessary to realize combined integral and lead control is illustrated by figure 109. The transfer-function of the resulting constant-resistance controller is given by Equation (409).

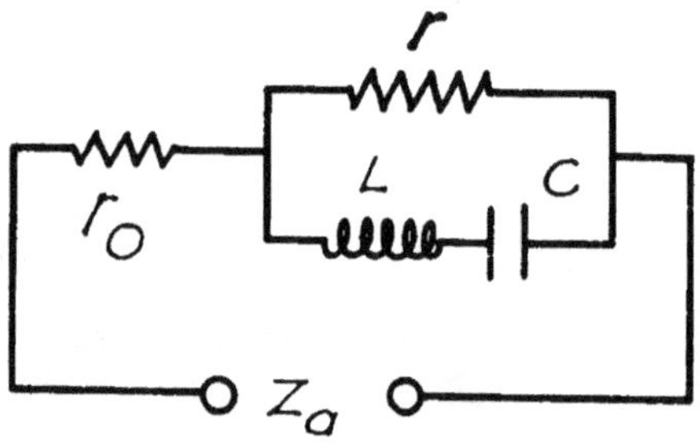

FIGURE 109

$$K_{Ldi}G_{Ldi}(s) = k_{L3}\,\frac{r_0 + r - 1}{r_0 + r + 1}\;\frac{LCs^2 + \dfrac{r(r_0-1)Cs}{r_0 + r - 1} + 1}{LCs^2 + \dfrac{r(r_0 + 1)Cs}{r_0 + r + 1} + 1} \tag{409}$$

The general form of the transfer-function of the combined lead-and integral-controller is given by Equation (322).

$$K_{di}G_{di}(s) = \frac{k_d k_i}{\alpha}\;\frac{\alpha\,\tau_d\tau_i s^2 + (\alpha\tau_d + \tau_i)\,s + 1}{a\,\tau_d\tau_i s^2 + (\tau_d + \alpha\tau_i)\,s + 1} \tag{322}$$

Equations (322) and (409) are equivalent provided that

$$\frac{k_d k_i}{\alpha} = k_{L3}\,\frac{r_0 + r - 1}{r_0 + r + 1} \tag{410}$$

$$\alpha\,\tau_d\tau_i = LC \tag{411}$$

$$\alpha\tau_d + \tau_i = \frac{r(r_0 - 1)C}{r_0 + r - 1} \tag{412}$$

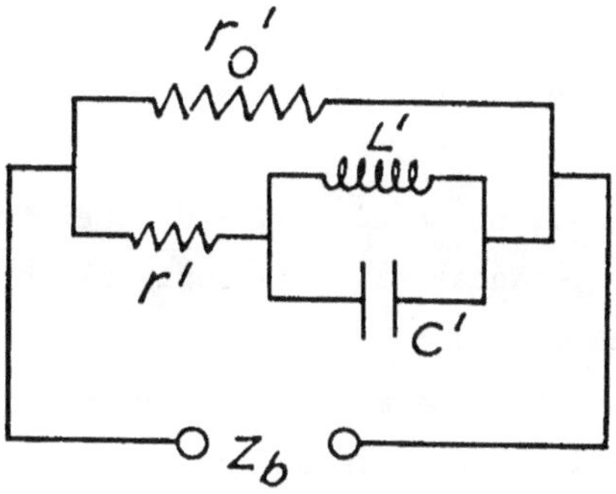

FIGURE 110

$$\tau_d + \alpha\tau_i = \frac{r(r_0 + 1)\,C}{r_0 + r + 1} \tag{413}$$

The values of the elements of z_a and the gain of the proportional amplifier can be found from the above four equations. Since there is one more unknown than there are defining equations, one element must be determined by other conditions. The resistance r_0 is the element whose value is most conveniently selected.

The impedance, z_b, that is the inverse of z_a is illustrated by figure 110.

The values of its components are given in terms of the components z_a , figure 109, by Equations (414) to (417).

$$r'_0 = \frac{1}{r_0} \tag{414}$$

$$r' = \frac{1}{r} \tag{415}$$

$$L' = C \tag{416}$$

$$C' = L \tag{417}$$

Both z_a, figure 109, and z_b, figure 110, employ non-dissipative inductors. However, the nature of the networks z_a and z_b are such that, provided the component values of the networks are readjusted and the resistive component of the inductor is not excessive, dissipative inductors can be used in a network that will be equivalent to the original. The necessary relations are not given here but are easily found by calculating and comparing the impedance functions of the two networks.

Matching Lead-Controller; Constant-Resistance Network

Chapter VI has shown that a device which has been termed a matching lead-controller can be very effective under certain circumstances in improving the transient performance of a servomechanism. The transfer-function of the matching lead-controller is given by Equation (249), and circuits that physically realize this transfer-function have been described in Chapters VI and IX. It is relevant to determine if this function can be physically realized by constant-resistance networks.

The locus of Equation (249) is not a circle, and, therefore, it cannot be realized by a lattice network whose component impedances have the forms illustrated by figure 102. The nature of the lattice impedances required to secure a controller whose transfer-function is equivalent to that given by Equation (249) can be determined however and the problem can be investigated of synthesizing these impedances. This problem is now undertaken.

Ithas been shown that Equation (385) is the expression for the transfer-function of a symmetrical, constant-resistance, lattice network. The impedance, z_a, is one of the lattice elements and the other is z_b, which is the inverse of z_a. Equation (385) can be solved for z_a.

$$z_a = \frac{1 - \frac{V_2}{V_1}(s)}{1 + \frac{V_2}{V_1}(s)} \tag{418}$$

The transfer-function $\frac{V_2}{V_1}(s)$ must be equal to the frequency-dependent portion of Equation (249) if the lattice network is to realize that function.

$$\frac{V_2}{V_1}(s) = \frac{s^2 + 2\zeta_c\omega_c s + \omega_c^2}{s^2 + 2\zeta_c n\omega_c s + a_c^2\omega_c^2} \tag{419}$$

The function z_a is found by substituting Equation (419) into (418) and simplifying the result.

$$z_a = \frac{2\zeta_c (n - 1)\omega_c s + (\alpha_c^2 - 1)\omega_c^2}{2s^2 + 2\zeta_c (n + 1)\omega_c s + (\alpha_c^2 + 1)\omega_c^2} \tag{420}$$

The problem is twofold; the first is to determine whether or not a function with the form of Equation (420) is realizable and the second is to find at least one network that physically realizes Equation (420) provided that such a network exists.

The existence of a physical network that will realize the function of Equation (420) may be determined by applying a criterion developed by Brune[7] in his paper, <u>Synthesis of a Finite Two-Terminal Network</u>. It may be shown that a physical network, that will exactly realize Equation (420), exists for certain values of the parameters ζ_c, α_c, n, and ω_c and fails to exist for other values. The criterion is given below and networks are developed in those instances in which they exist. Brune's method is followed throughout, and reference is made to his paper for proof of certain relations.

For simplicity, Equation (420) may be written

$$z_a = \frac{a_0 + a_1 s}{b_0 + b_1 s + b_2 s^2} \tag{421}$$

in which

$$a_0 = (\alpha_c^2 - 1)\omega_c^2$$

$$a_1 = 2\zeta_c(n - 1)\omega_c$$

$$b_0 = (\alpha_c^2 + 1)\omega_c^2$$

$$b_1 = 2\zeta_c (n + 1)\omega_c$$

$$b_2 = 2$$

The inverse of z_a is

$$\frac{1}{z_a} = \frac{b_0 + b_1 s + b_2 s^2}{a_0 + a_1 s} \tag{422}$$

The right hand side of Equation (422) can be separated into two parts by dividing the numerator by the denominator.

$$\frac{1}{z_a} = \frac{b_2 s}{a_1} + \frac{b_0 + (b_1 - \frac{a_0}{a_1} b_2)s}{a_0 + a_1 s} \tag{423}$$

The first term in (423), $\frac{b_2}{a_1} s$, is the admittance of a capacitor whose value is

$$\frac{b_2}{a_1} = C_1 \tag{424}$$

The second term in (423) is the expression for the inverse of an impedance that can be denoted by z_1

$$z_1 = \frac{a_0 + a_1 s}{b_0 + (b_1 - \frac{a_0}{a_1} b_2)s} \tag{425}$$

Thus z_a may be regarded as the parallel combination of a capacitance whose value is now known and an impedance, z_1, of simpler form than that of the original impedance z_a.

Let $b_1 - \frac{a_0}{a_1} b_2 = d$ (426)

Then $z_1 = \frac{a_0 + a_1 s}{b_0 + ds}$ (427)

The physical realizability and nature of the resulting network depends upon the sign of the quantity, d. The three possible cases, d greater than, equal to, or less than zero are discussed individually below.

Case 1. $d > 0$

The physical significance of a positive value of d is found by expanding Equation (426).

$$d = b_1 - \frac{a_0}{a_1} b_2 > 0 \qquad (426a)$$

$$2\zeta_c (n + 1)\omega_c - \frac{(\alpha_c^2 - 1)\,\omega_c}{\zeta_c (n - 1)} > 0 \qquad (428)$$

$$2\zeta_c^2 (n^2 \quad 1) > \alpha_c^2 - 1 \qquad (429)$$

The physical network required to realize the function given by Equation (427) depends upon the sign of the expression $(a_0 d - a_1 b_0)$. If this quantity is positive, the network has the form illustrated by figure 111. The impedance, z_1', of this network is

$$z_1' = \frac{R_1 + R_c + R_1 R_c C_2 s}{1 + R_c C_2 s} \qquad (430)$$

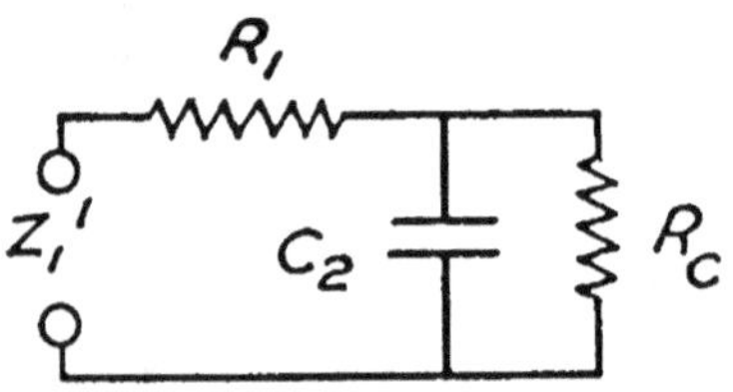

FIGURE 111

Equation (430) is similar to Equation (427) and is equivalent if the following relations hold.

$$R_1 + R_c = \frac{a_0}{b_0} \qquad (431)$$

$$R_c C_2 = \frac{d}{b_0} \qquad (432)$$

$$R_1 R_c C_2 = \frac{a_1}{b_0} \qquad (433)$$

The above three equations can be solved for the network elements, R_1, R_c, and C_2.

$$R_1 = \frac{a_1}{d} \qquad (434)$$

$$R_c = \frac{a_0}{b_0} - \frac{a_1}{d} = \frac{a_0 d - a_1 b_0}{b_0 d} \qquad (435)$$

$$C_2 = \frac{d^2}{a_0 d - a_1 b_0} \qquad (436)$$

Equations (435) and (436) show that R_c and C_2 have positive values only if $(a_0d - a_1b_0)$ is positive. Therefore, Equation (427) is not realizable by the network of figure 111 unless $(a_0d-a_1b_0)$ is positive.

If unrealizable by the network of figure 111, Equation (427) may be realized by the network illustrated by figure 112.

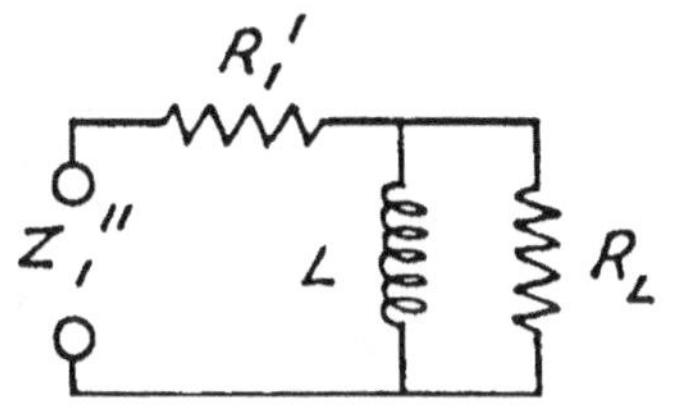

FIGURE 112

The impedance, z_1'', of this network is

$$z_1'' = \frac{R_1' + \dfrac{R_1' + R_2}{R_2} Ls}{1 + \dfrac{Ls}{R_2}} \qquad (437)$$

Equations (427) and (437) are equivalent provided that

$$R_1' = \frac{a_0}{b_0} \qquad (438)$$

$$R_L = \frac{a_1b_0 - a_0d}{b_0 d} \qquad (439)$$

$$L = \frac{a_1b_0 - a_0d}{b_0^2} \qquad (440)$$

It is evident from Equation (439) and (440) that the elements R_L and L, of figure 112 are positive only if $(a_0d - a_1b_0)$ is negative.

The results of the case in which d is positive are summarized in figure 113.

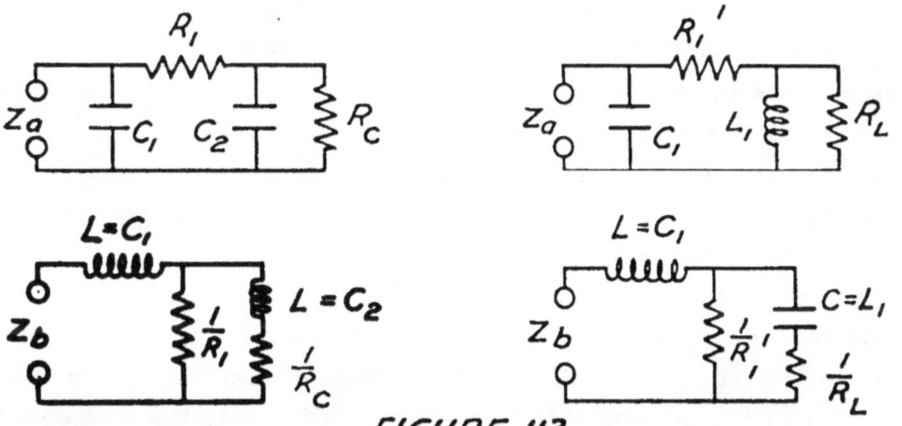

FIGURE 113

$$C_1 = \frac{b_2}{a_1}$$

$$R_1 = \frac{a_1}{d}$$

$$C_2 = \frac{d^2}{a_0 d - a_1 b_0}$$

$$R_c = \frac{a_0 d - a_1 b_0}{b_0 d}$$

$$C_1 = \frac{b_2}{a_1}$$

$$R_1' = \frac{a_0}{b_0}$$

$$L_1 = \frac{a_1 b_0 - a_0 d}{b_0^2}$$

$$R_1 = \frac{a_1 b_0 - a_0 d}{b_0 d}$$

Case 2. d = 0

For the case in which d = 0, the expression for z_a, Equation (427), reduces to

$$z_1 = \frac{a_0}{b_0} + \frac{a_1}{b_0} s \tag{441}$$

Equation (441) is the impedance of a resistance and inductance in series. The networks, z_a and z_b, are shown in figure 114.

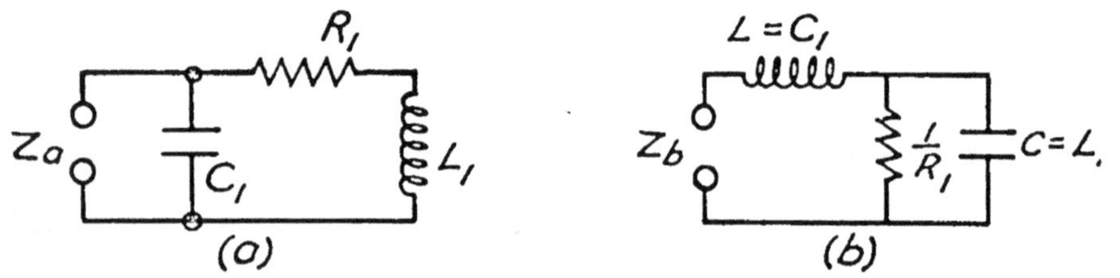

$$C_1 = \frac{b_2}{a_1}$$

$$R_1 = \frac{a_0}{b_0}$$

$$L = \frac{a_1}{b_0}$$

FIGURE 114

Case 3. d > 0

If the quantity d is less than zero, the impedance, z_a, cannot be physically realized by a constant-resistance network. For proof of this, reference is made to Brune's paper. The reason for this non-realizability is that the transfer-function, Equation (249), for certain frequencies has an absolute magnitude greater than one if the quantity d is less than zero. However, if the right hand side of Equation (249) were multiplied by a constant whose magnitude is sufficiently smaller than unity that the magnitude of $\frac{V_2}{V_1}(s)$ did not exceed unity, the new expression can be realized by a constant-resistance network.

The effect of multiplying by the constant is not to change the nature and shape of the function in the frequency scale but to reduce the output level so that the magnitude of the output voltage never exceeds the magnitude of the input voltage.

This is a perfectly proper procedure because the output level for proper servo operation can be retained by correspondingly increasing the gain of the amplifiers preceding or following the constant-resistance network.

Bridged-T Equivalent of the Lattice

Those compensating circuits in lattice form whose transfer-loci are circles can be converted into an equivalent bridged-T network. Since the bridged-T network is an unbalanced structure, in contrast with a lattice network that is a balanced structure, the bridged-T is more adaptable to incorporation in electronic circuits where it is necessary to preserve a potential reference point throughout the circuit. The conversion from the lattice to the bridged-T network is explained in detail by Guillemin.* It is outlined briefly below. The circuit of a bridged-T network is illustrated by figure 115.

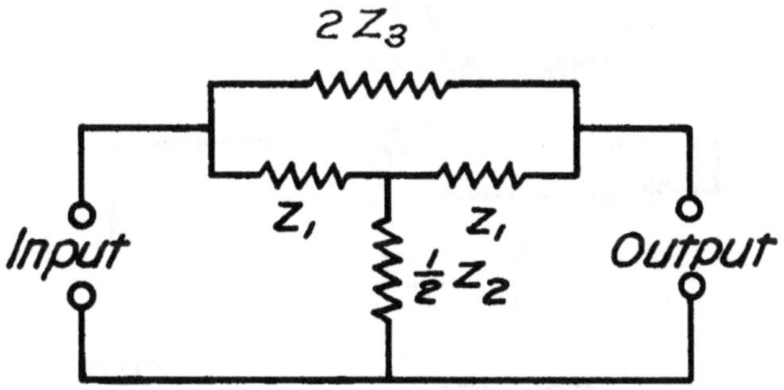

FIGURE 115

The conversion from the lattice to the bridged-T utilizes the fact that in all cases where the locus of the lattice impedance, z_b, is a circle, this impedance comprises a parallel combination of a resistance and a second impedance. (See figure 102b). Therefore, the expression for the inverse of z_b can be written

$$\frac{1}{z_b} = \frac{1}{z_b'} + \frac{1}{r_0'} \qquad (442)$$

in which r_0' is the parallel resistance and z_b' the parallel impedance. The conditions for equivalence between the lattice of figure 101 and the bridged-T of figure 115 follow.

$$z_1 = r_0' \qquad (443)$$

$$z_2 = \frac{1}{z_b'} + \frac{1 - r_0'^2}{r_0'} \qquad (444)$$

$$z_3 = z_b' \qquad (445)$$

As an example, the bridged-T, constant-resistance network with a transfer-function equivalent to that of an integral-controller is depicted in figure 116. The circuit elements, L', r' r_0', r, and C are expressed in terms of the integral-controller constants, α_i and τ_i by Equations (392) and (393) and the relations in figure 105b.

The subject of equivalent circuits is a broad one and it is the purpose of this paper to do no more than indicate the possibilities. For further information, reference is made to texts and articles on network theory.

---

* See the second volume of reference 10

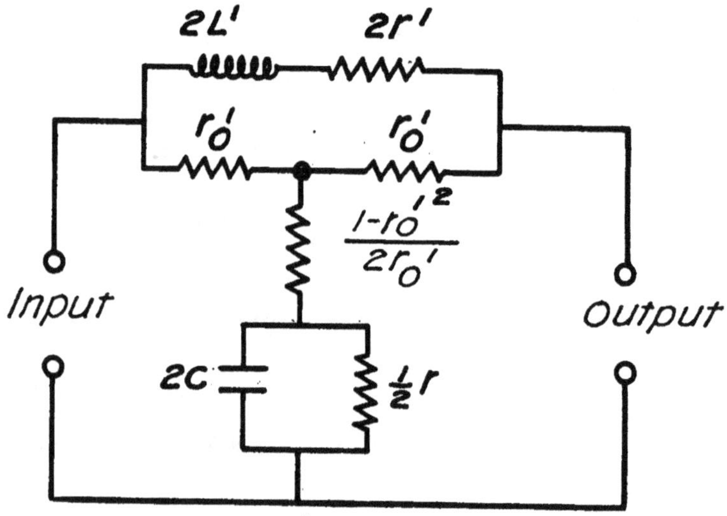

FIGURE 116

SUGGESTIONS FOR FURTHER WORK

Techniques of servomechanism design have been developed in this paper that are based primarily upon the properties of the servo transfer-function and its locus. The efficacy of these techniques has been demonstrated in the solution of the following problems: (1) The selection of basic servo systems for particular applications. (2) The proper adjustment of these systems to secure optimum performance. (3) The design of compensating functions to improve the performance of the basic systems. (4) The synthesis of various types of physical devices to realize these compensating functions. There are several other problems of importance in the design of linear servomechanisms and also alternative methods of approaching certain problems discussed herein that have not been treated in the paper. The basic theory of the properties of transfer-functions and their loci as developed in this paper is also an aid to the solution of these problems. It is of interest, therefore, to outline certain of these problems and briefly indicate possible approaches to their solution.

Servomechanism Compensation by Means of Paralleled Devices

All of the compensating circuits described in this paper have been devices that were cascaded with the basic portion of the servomechanism as illustrated by figure 4. Another possible method of compensation is that of paralleling a properly designed device with the element of the servomechanism whose performance requires improvement. This element is usually the servo motor or the servo motor and the device controlling it; it could, however, be any component of the system. The compensating device should be so designed that the parallel combination is characterized by damping ratios and characteristic frequencies that are necessary to secure the servo performance required. For example, if the servo motor requires improvement, the basic circuit of the compensating device might be such that the parallel combination tends to approximate an ideal servo motor of the integrating type. (See page 42). The methods of analysis explained in the paper, with appropriate modifications, should be of considerable assistance in the design of such devices.

Design of Sense-Detecting Rectifiers

In many servo systems, the signal proportional to the servo error may be in the form of a modulated carrier wave. The reason for this is that the output of certain instruments widely employed to compare the position of the servo output and the servo input is in the form of a modulated carrier wave, and also because signals of this nature are more easily transmitted. In this case it is necessary to demodulate these signals at some point of the system in such a fashion that the output of the demodulator is a suitable function of both the magnitude and the sense of the error between the servo output and input. It is frequently convenient to perform the demodulation by means of vacuum-tube circuits and in this case the demodulator is known as a <u>sense-detecting rectifier</u>. The sense-detecting rectifier is a vital element in those systems of which it is a part, and it is important that it be designed to yield optimum performance. The output of the sense-detecting rectifier generally must be filtered by some means in order to attenuate those signal components present in its output that are multiples of the carrier frequency. If these high frequency signal components are insufficiently attenuated they are likely to overload following stages. On the other hand, if they are filtered improperly or excessively, the filter will introduce a lag into the system that will damage the performance of the servo. It is important, therefore, that the design of the rectifier circuit and the filter take these points into consideration.

It is difficult to calculate the response characteristics of a sense-detecting rectifier because of the non-linear properties of that type of device, but it is relatively easy to obtain these characteristics by measurement. A study of this element of the system and the compilation of its characteristics in the form of curves of the magnitude and phase of its transfer-function, or parametric graphs of the transfer-locus, would be most useful. Some of this work has been done by the author.

The problem of the design of the sense-detecting rectifier is especially important if phase-lead-controllers are employed in the system. If these devices follow the demodulator, they generally amplify the high frequency signal components produced by the rectifier and the result may be serious overloading of subsequent stages. The lead-controller can be designed to have a response similar to that illustrated by figure 65, and the response of the system at the carrier frequency can be made small. However, if the carrier frequency is too low, the characteristics of the controller in the positive phase-shift region are seriously impaired.

A careful study of this problem should be of considerable value. It should be possible to specify the lower limit of carrier frequency that can be employed with given types of lead-controllers.

<u>Design of Carrier-Frequency Compensating Devices</u>

All of the compensating devices that have been described in this paper have operated upon the error signal directly. In a servomechanism employing carrier translation of the error, these compensating devices would properly follow the sense-detecting rectifier and operate upon the demodulated output. It is possible, however, to design circuits that operate with the same effect upon the modulated wave itself. That is, it is possible to design circuits that, inserted in the carrier channel, have an equivalent effect upon the system transfer-function as the circuits hitherto described that are inserted in the channel of the demodulated signal, <u>provided</u> that the carrier frequency is not too low. Such devices are very useful in certain applications.

There are several means of determining these carrier-frequency circuits. One method is as follows: Take the low-frequency circuit that it is desired to simulate and (1) replace each inductor present by a series-resonant circuit, tuned to the carrier frequency and including an inductor whose value is one-half that of the corresponding inductance in the low-frequency circuit; (2) replace each capacitor present in the low-frequency circuit by a parallel-resonant circuit tuned for the carrier frequency and including a capacitor whose value is one-half that of the corresponding capacitance of the low-frequency circuit; (3) maintain invariant the resistor values of the low-frequency circuit.

The merit of this particular circuit depends in general upon securing inductors with high time constant. In certain instances carrier-frequency equivalent circuits may be secured that do not require inductors. For example, bridge-T circuits, twin-T circuits, and Wien bridge circuits can be used to realize lead-control compensation.

Circuits of this nature can be found that realize integral-control, basic lead-control, compound lead-control, or matching lead-control. The circuits obtained may be multiple stages and require buffer amplifiers between successive stages or be constant-resistance networks with the characteristics described in the last chapter. There is room for considerable work on the problem of designing these circuits. The effect of shifts in the carrier frequency, which appears to be a fundamental limitation, needs to be investigated as well as the effect of dissipation in the inductors.

The preceding examples are a few problems which are solvable by the methods previously developed. These methods should be sufficiently general to apply to other problems that will undoubtedly occur to the reader.

BIBLIOGRAPHY

1. "Automatic Pilot", Engineer, Vol. 159, pp. 120-2, February 1, 1935.
2. Black, H. S. "Stabilized Feed-Back Amplifiers", Bell System Technical Journal, Vol. 13, pp. 1-18, January, 1934.
3. Black, H. S. "Wave Translation System", U. S. Patent No. 2,102, 671, December 21, 1937.
4. Bode, H. W. "Relations Between Attenuation and Phase in Feedback Amplifier Design", Bell System Technical Journal, Vol. 19, pp. 421-54, July, 1940.
5. Bode, H. W. "Amplifier", U. S. Patent No. 2, 123, 178, July 12, 1938.
6. Brown, G. S. "Behavior and Design of Servomechanisms", Printed November, 1940, under the auspices of the Fire Control Committee (Sec. D-2) of NDRC.
7. Brune, O. "Synthesis of a Finite Two-Terminal Network Whose Driving Point Impedance Is a Prescribed Function of Frequency", Journal of Mathematics and Physics, Vol. 10, pp. 191-236, 1930-31.
8. Callender, Hartree, and Porter "Time Lag in a Control System", Roy. Soc. Phil. Trans., Vol. 235A, pp. 415-44, July 21, 1936.
9. Gardner, M. F., and Barnes, J. L. Transients in Linear Systems, John Wiley and Sons, New York, 1942.
10. Guillemin, E. A. Communication Networks, Vols. 1 and 2, John Wiley and Sons, New York, 1935.
11. Gulliksen, F. H. "Recent Developments in Electronic Devices for Industrial Control", A.I.E.E. Paper No. 33-24, 1933.
12. Haigler, E. D. "Application of Temperature Control", Trans. A.S.M.E., Vol. 60, pp. 633-40, 1938.
13. Harris, H. "The Analysis and Design of Servomechanisms", Printed 1942 under the auspices of the Fire Control Committee (Sec. D-2) of NDRC., OSRD No. 454.
14. Hazen, H. L. "Theory of Servomechanisms", Journal of the Franklin Institute, Vol. 218, No. 3, September, 1934.
15. Mason, C. E. and Philbrick, G. A. "Automatic Control in the Presence of Process Lags", Trans. A.S.M.E., Vol. 62, pp, 295-308, 1940.
16. McCrea, H. A. "Automatic Control of Frequency and Load", Gen. Elec. Rev., Vol. 32, No. 6, pp. 309-13, June, 1929.
17. Minorsky, N. "Automatic Steering Tests", Journ. Am. Soc. Naval Engrs., Vol. 42, No. 2, p. 285, May, 1932.
18. Minorsky, N. "Directional Stability of Automatically Steered Bodies", Jour. Am. Soc. Naval Engrs., Vol. 34, No. 2, pp. 280-309, May, 1922.

19. Mitereff, S. D. "Principles Underlying the Rational Solution of Automatic Control Problems", A.S.M.E. Trans., Vol. 57, pp. 156-63, May, 1935.

20. Nichols, N. B. and Marcy, H. T. Two separate papers prepared and privately distributed by the Servomechanisms Laboratory at M.I.T., 1942.

21. Nyquist, H. "Regeneration Theory", Bell System Tech. Jour., pp. 126-47, January, 1932.

22. Sperry, E. A., Jr. "Description of the Sperry Automatic Pilot", Aviation Engr., pp. 16-18, January, 1932.

23. Thompsen, J. W. "Automatic Control Applications Increase as Refineries find Operation Economies", Nat. Pet. News, Vol. 27, p. 26, March 27, 1933.

24. Weiss, H. K. "Constant Speed Control Theory", Jour. of the Aero Sciences, Vol. 6, No. 4, February, 1939.

25. Werey, R. B. "Instrumentation and Automatic Control in the Oil Refining Industry", The Brown Instrument Company, Philadelphia, Pennsylvania, 1941.

26. Whittaker, E. T., and Watson, G. N. Modern Analysis, Macmillan Company, New York, 1935.

27. Zabel, R. M. "The Use of Thyratrons for Temperature Control", R. S. I., Vol. 28, January, 1934.

28. Ziegler, J. G. and Nichols, N. B. "Optimum Settings for Automatic Controllers", Trans. A.S.M.E., preprint, December, 1941.

29. Ziegler, J. G. and Nichols, N. B. "Process Lags in Automatic Control Circuits", Trans. A.S.M.E., preprint, October, 1942.

30. Lee, Y. W. "Synthesis of Electric Networks by Fourier Transformation of Laguerre's Functions". Jour. of Mathematics and Physics, Vol. 11, June, 1932, pp. 83-113.

31. Bomberger, D. C. and Weber, B. T., "Stabilization of Servomechanisms". Bell Telephone Laboratories, December 10, 1941, M.M-41-110-52.

www.ingramcontent.com/pod-product-compliance
Lightning Source LLC
Chambersburg PA
CBHW080546220326
41599CB00032B/6382